COLOR ATLAS

OF

Microbiology

R. J. OLDS

PhD, BVSc, DpBact
Lecturer
Department of Pathology,
University of Cambridge

YEAR BOOK
MEDICAL PUBLISHERS, INC

35 E. WACKER DRIVE, CHICAGO

Copyright © R J Olds 1975
This book is copyrighted in England and may not
be reproduced by any means in whole or part.
Distributed in Continental North, South, and Central America,
Hawaii, Puerto Rico, and the Philippines by
Year Book Medical Publishers, Inc
By arrangement with
Wolfe Publishing Ltd
Library of Congress Catalog Card Number : 70-165961
International Standard Book Number : 0-8151-6542-0
Printed by Smeets-Weert Holland

Year Book Color Atlas Series
Series Editor
G Barry Carruthers MD (Lond)

Other books in this series already published :
Color atlas of General Pathology
Color atlas of Oro-Facial Diseases
Color atlas of Ophthalmological Diagnosis
Color atlas of Renal Diseases
Color atlas of Venereology
Color atlas of Dermatology
Color atlas of Infectious Diseases
Color atlas of E.N.T. Diagnosis
Color atlas of Rheumatology

Further titles now in preparation :
Color atlas of Pediatrics
Color atlas of Forensic Pathology
Color atlas of Gynecology
Color atlas of Physical Signs in Clinical Medicine
Color atlas of Tropical Medicine & Parasitology
Color atlas of Histology
Color atlas of Gastro-Intestinal Endoscopy
Color atlas of Orthopedics
Color atlas of the Liver
Color atlas of Virology
Color atlas of Dental Surgery
Color atlas of Cardiology
Color atlas of Respiratory Diseases
Color atlas of Endocrinology
Color atlas of Surgical Diagnosis

Acknowledgements

Three of the photographs are from transparencies generously provided by others for the purpose. The photograph for **168** was taken by Dr P Walker and that for **376** by Dr Shireen Chantler; these two transparencies were supplied by Dr Irene Batty. The transparency for **172** was supplied by Dr A E Wilkinson.

Many people have kindly contributed material for me to photograph. Eighty of the photographs are of primary plate cultures and other materials collected by my friends in the Public Health Laboratory Service at Cambridge. For this I am particularly indebted to Dr G R E Naylor and Dr Joan Boissard, each of whom also made many helpful suggestions during the preparation of the text, and to Dr H W K Fell, whose contributions appear in both bacterial and fungal chapters. Dr J Marks, Tuberculosis Reference Laboratory, Cardiff, contributed specimens for 24 photographs of mycobacteria. Specimens for 38 photographs of fungi were obtained from workers at the Mycological Reference Laboratory, London, through the courtesy of the late Dr I G Murray and Dr D W R Mackenzie; Miss Christine Philpot and Mr A G Proctor have been most helpful. Fifty-two specimens were prepared by our technical staff, Mr A H Westley, Mrs Lesley Allinson and Miss Julie Matthews, and 15 by undergraduates in this department. Most of the media were prepared by Mr J F Stevenson and his staff. Each of these groups have contributed so much that itemised acknowledgement would be excessively detailed, yet quite inadequate to express the value of their association.

Material for photography was supplied also by Dr T J Alexander and staff for **158**, **159**, **163**, **176**, **183** and **187**; Professor H R Carne for **28**, **292**, **312**, and **378-382**; Dr L M Dowsett for **132**; Dr D Franks for **362**; Professor H M Gilles for **175**, **184** and **186**; Dr M H Gleeson-White and staff for **17**, **18**, **108**, **133**, **138** and **154**; Mr S A Hall for **119–122** and **297**; Mr R G Hirst for **117**, **338**, **346** and **351**; Dr C E Hormaeche for **372–374**; Mrs Yvonne Lawson for **358**; Dr S W B Newsom for **294**, **295** and **298**; Dr T A Roberts for **394**; Dr P Whittlestone for **125**, **126**, **182** and **369**; Dr A T Willis for **54**. The specimens shown in **136**, **153**, **257** and **383–385** were from this Department's permanent collection of teaching materials.

I am indebted to Dr G R E Naylor and the Public Health Laboratory Service for permission to publish **272** and **273**, and to Dr R M Fry for his critical comments on much of the text. My wife's help has taken diverse forms — technical, clerical and critical. The publishers have been most helpful and patient.

It is a pleasure to acknowledge the assistance of all of these persons and to record my thanks for their contributions.

To my parents

Contents

Preface: Materials and Methods

Most of the procedures were those of Cruickshank (1969). Hartley's digest broth was used as the basis for digest agar and blood agar. Media were designated A or B according to whether they were made in the Public Health Laboratory or in the Department of Pathology, Cambridge. Some slight colonial differences were found consistently on media made in the two laboratories. Colonies essentially similar to those on blood agar A can be grown on blood agar made with Columbia agar (Baltimore Biological Laboratory; Oxoid Limited). Methods other than those of Cruickshank (1969) are mentioned in the following paragraphs.

The procedures of Cowan and Steel (1965) were used for the preparation of potato medium, and for tests for decarboxylases, Frazier's method for hydrolysis of gelatin, Hugh and Leifsen's method for oxidation or fermentation of carbohydrates, lecithinase, levan and dextran formation, nitratase, phenylalanine, phosphatase and starch hydrolysis tests.

Media and methods for the growth of mycobacteria were described by Marks (1972), and for fungi by Collins and Lyne (1970).

Fluid thioglycollate medium, TCBS agar and triple sugar iron agar were produced by Difco Laboratories; bacitracin and optochin sensitivity discs, multodiscs and sensitest agar by Oxoid Limited.

For other materials and methods see: Anderson *et al* (1931) for McLeod's heated blood tellurite agar; Bühlmann *et al* (1961) for the decomposition of acetamide test; Crowle (1961) for the Amido Schwarz stain; Gershman (1963) for the motility-sulphide medium; Gridley (1953) for his fungal stain; Lacey (1954) for his pertussis medium with antibiotics; Lautrop (1960) for his urease test; Mackie and McCartney (1953) for Kirkpatrick's flagella stain; Preston and Maitland (1952) for their flagella stain; Preston and Morrell (1962) for their Gram stain; Walker *et al* (1971) for fluorescent antibody staining; Wheeler *et al* (1965) for the TRIFF stain; and Whittlestone (1969) for mycoplasma media.

Glossary: **Some Common Synonyms**

The first name listed for each organism is used in this book. The other names are cross-referenced in the index.

Acinetobacter anitratum: Achromobacter anitratus, Cytophaga anitratum, Moraxella glucidolytica.

Actinobacillus aprophilus: Haemophilus aprophilus.

Bordetella bronchiseptica: Alcaligenes bronchisepticus, Brucella bronchiseptica.

Bordetella parapertussis: Acinetobacter parapertussis.

Campylobacter fetus: Vibrio fetus.

Citrobacter ballerup: Ballerup-Bethesda Group, *Salmonella ballerup.*

Citrobacter freundii: Escherichia freundii, Colloides anoxydana.

Clostridium chauvoei: Clostridium feseri.

Clostridium welchii: Clostridium perfringens.

Corynebacterium haemolyticum: Corynebacterium pyogenes var. *humanis.*

Enterobacter aerogenes: Aerobacter aerogenes (motile strains).

Enterobacter cloacae: Aerobacter cloacae, Cloaca cloacae.

Fusobacterium fusiforme: Fusiformis fusiformis, Bacteroides fusiformis.

Gonococcus : *Neisseria gonorrhoeae.*

Klebsiella aerogenes: Aerobacter aerogenes (non-motile strains).

Klebsiella pneumoniae: Friedländer's bacillus.

Meningococcus : *Neisseria meningitidis.*

Micropolyspora faeni: Thermopolyspora polyspora.

Moraxella: includes *Mima polymorpha* var. *oxidans.*

Mycobacterium intracellulare: Battey bacillus.

Neisseria mucosa: Diplococcus mucosus.

Nocardia madurae: Actinomadura madurae, Streptomyces madurae.

Paracoccidioides brasiliensis: Blastomyces brasiliensis.

Pasteurella septica: Pasteurella multocida.

Phialophora pedrosoi: Fonsecaea pedrosoi, Hormodendrum pedrosoi.

Pneumococcus : *Diplococcus pneumoniae, Streptococcus pneumoniae.*

Pseudomonas aeruginosa: Pseudomonas pyocyanea.

Pseudomonas mallei: Acinetobacter mallei, Loefflerella mallei, Malleomyces mallei, Pfeifferella mallei, the glanders bacillus.

cont.

Pseudomonas pseudomallei: Loefflerella whitmori, Malleomyces pseudomallei, Pfeifferella whitmori, Whitmore's bacillus.

Salmonella arizona: Arizona group, *Paracolobactrum arizona Salmonella* subgenus III.

Salmonella paratyphi B: Salmonella schottmülleri.

Staphylococcus albus: largely synonymous with *Staphylococcus epidermidis.*

Staphylococcus aureus: Staphylococcus pyogenes.

Streptobacillus moniliformis: Actinobacillus muris, Streptothrix muris ratti.

Streptococcus, Group A : *Streptococcus pyogenes.*

Group H : *Streptococcus sanguis.*

Group K : *Streptococcus hominis, Streptococcus salivarius.*

Thermoactinomyces vulgaris: Micromonospora vulgaris.

Trichophyton sulphureum: Trichophyton tonsurans var. *sulphureum.*

Vibrio cholerae: Vibrio comma.

Yersinia pestis: Pasteurella pestis, the plague bacillus.

Yersinia pseudotuberculosis: Pasteurella pseudotuberculosis.

Introduction

This atlas is intended primarily for students of microbiology in the medical and paramedical sciences. It should also meet some needs of laboratory workers and of student technicians in microbiology. Its purpose is to illustrate the appearances of bacteria and fungi as they are seen at the laboratory bench.

Each photograph is accompanied by essential explanatory text. The clinical significance of the result illustrated is seldom mentioned. If it is not obvious it can be found in standard textbooks to which it is assumed that the student has access. Each caption includes a statement of the magnification represented by the printed photograph. Where possible these magnifications are comparable with those most used in the laboratory. Thus, photomicrographs of bacteria are reproduced at x1000, colonies of bacteria at x6 – a magnification readily obtained with a hand lens, and petri dish and fungal tube cultures at about x1.

Standard methods were used for photomicrography. For macro-photography one or more of the methods illustrated in **1** to **4** were used. Details of media and other materials and methods will be found in the preface.

Most of the commonly encountered bacteria and fungi are illustrated. Some of the less common organisms are shown as well. These have been selected for one or more of the following reasons: they may be dangerous pathogens, in which case it seems better to expose the student to a photograph then to a live culture; they may be sufficiently similar to a dangerous pathogen to represent a better organism for student study than the pathogen itself; they may illustrate some germinal point in the development of microbiology, such as the L-forms of *Streptobacillus moniliformis.*

Virtually all features which may be used in the characterisation of a micro-organism are likely to vary with the medium on which it is grown. As media become more standardised this source of variation between laboratories is becoming reduced. Within any one laboratory standardisa-tion of media by strict quality control and evaluation is essential for the skilled examination of cultures. Apart from variability resulting from variation in culture media, one can expect differences between strains within a bacterial or fungal species. The art of the bacteriologist or the mycologist is his recognition at a glance of features which may not be described adequately by the written word. It is hoped that this colour atlas has a part to play in the development of this art.

Clues to the source and methods of spread of infection in a community may come from any quarter. Probably the most fruitful source is colony morphology, because it is promptly available and because the colony has manifold features, any one of which may provide the clue. Other sources are unusual biochemical properties or patterns of antibiotic sensitivity. Any one of these may be just as useful as an epidemiological marker as phage-, bacteriocin- or sero-type which are deliberately and painstakingly sought for the purpose.

Key

= lamp with hemispherical reflector

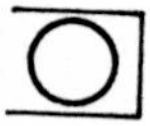

= lamp with cylindrical reflector

= subject

= camera

= black background

= white background or reflector

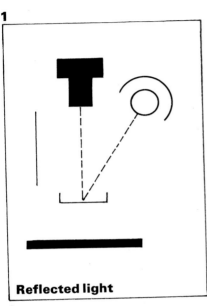

Reflected light

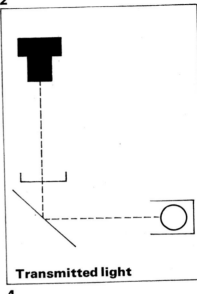

Transmitted light

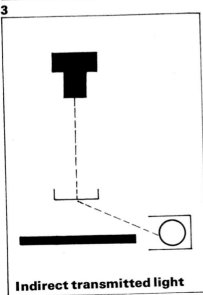

Indirect transmitted light

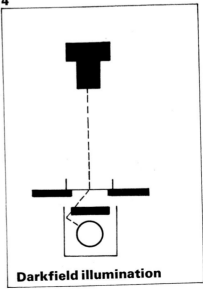

Darkfield illumination

11

Bacteria: Macroscopic Appearances

This section depicts a range of colony morphology on a fairly standard set of media. The object is to give guidance in examining colonies and to illustrate the main features, rather than to depict standards by which the identity of an unknown can be established. Each bacteriologist must establish his own standards for his own media. The photographs should help him do this, but they cannot render the process unnecessary.

The photographs are intended to replace the detailed verbal descriptions of colonies found in many textbooks. Therefore the caption does not include a long description. Sometimes the characteristics of a colony can be stated briefly by noting its resemblance to some well-known object, as for example, when a pneumococcal colony is described as a 'draughtsman' or 'chequer'. These terms have proved very useful to practising bacteriologists, and a number of them are used in this section.

There is no adequate photographic substitute for the actual examination of colonies if the student is to appreciate the three-dimensional structure of the colony. However, something of this structure will be appreciated if he scans the photograph from top to bottom, and notices how the incident light illuminates colonies at the top of the picture where the light is almost over the colony, and then examines colonies near the bottom where the angle of illumination is flatter. Thus, although the plate remains still in the photograph, by viewing various colonies, he can obtain some impression of what he sees when looking at one colony and moving the plate.

Some preliminary examples
It may be useful to study a few selected examples first. In examining a photograph, particular note should be made of how the lamp is imaged on the surface of the colony : a smooth colony will produce a sharp image (**5**) ; if the surface is matt, the image will be less sharp (**10**) ; if the surface is quite rough, no image will be seen (**48**). The image of the lamp also gives some idea of the elevation of the colony : in a smooth

convex colony an undistorted view of the lamp will be seen (**111**) ; a conical colony will produce a triangular image (**26**) ; if the elevation is irregular, the image will be distorted accordingly (**86**).

The edge of the colony may be entire (**5**) or it may have some degree of irregularity. The most irregular edges are found on rough-surfaced colonies (**45**). How the edge of the colony slopes down to the medium, i.e., the elevation at the edge, is a most distinctive feature of some colonies. Compare for example the 'shorelines' of the three colony types in (**72**.)

By transmitted light a colony may appear transparent, i.e. water clear (**106**), translucent or opaque (small and large colonies in **66** respectively). Colony colour should not be assessed by transmitted light except when specifically looking for the distinctive colours shown by many species by indirect transmitted light (**66**). Otherwise, the colour of a colony is best appreciated by diffuse reflected light, but because diffuse lighting understates colony shape it has not been used in this atlas.

Most colonies have a butyrous (buttery) consistency ; this is appreciated by manipulating the growth with a wire. Some colonies fracture when touched, others stick to the wire (**22**) or appear like gum (**91**).

5 *Colonies of two species of staphylococci* The larger are *S. aureus*; note that the distinctive golden pigment is most intense in the centre of the colony. The smaller colonies are *S. albus*. Not all *albus* strains have such small colonies. Both colony types are low-convex, smooth and opaque. (*Digest agar B, 24 hours at 37°C; reflected light, x6.*)

6 *Staphylococci and diphtheroids cultured from a nasal swab* The numerous small colonies are typical diphtheroids of the nose. There are two types of staphylococcal colony: the smaller, haemolytic colonies inhibit growth of the diphtheroid; the larger, non-haemolytic ones do not. Inhibition is caused by a bacteriocin, which is produced only by *Staphylococcus aureus* of phage type 71. This phage type is implicated in many cases of impetigo. (*Blood agar A, overnight at 37°C; reflected light, x6.*)

7 *Three colonies of micrococci* The micrococci are frequently encountered as aerial contaminants. Many strains produce yellow or pink pigments. Smears frequently reveal an arrangement in groups of four cells; this applied to all three colonies shown here (see **130**). The yellow colonies have often been called *Staphylococcus citreus* or *Micrococcus luteus*. (*Digest agar B, 48 hours at 37°C; reflected light, x6.*)

8 *Colonies of alpha-haemolytic streptococci* They are surrounded by a zone of grey-green discoloured haemoglobin. The pigment remains fixed in the vicinity of the colony. Alpha-haemolytic streptococci are very common in the upper respiratory tract. They are often called *S. viridans*, but it is clear that this is a heterogeneous group of colonial, biochemical and serological types, for which a Linean binomial is inappropriate. (*Blood agar A, 18 hours at 37°C; reflected + transmitted light, x6.*)

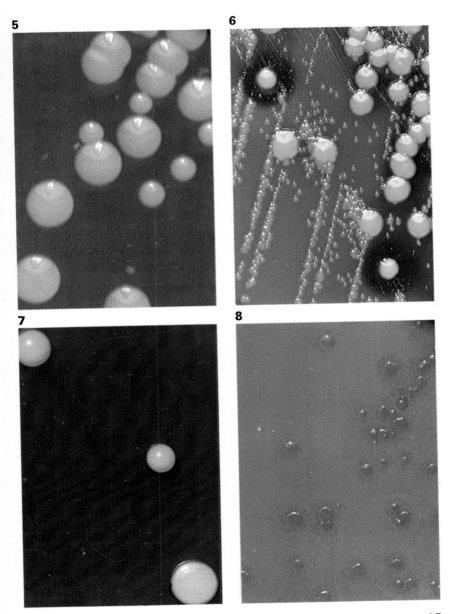

9 & 10 *Streptococcus pyogenes colonies* These show haemolytic zones of different diameters. Strains which produce large zones of haemolysis (**9**) are readily recognized. Strains of S. pyogenes which have virtually no haemolytic zone (**10**) are not uncommon. (*Blood agar A, 18 hours at 37°C; reflected + transmitted light, x6.*)

11 *Streptococci of two Lancefield groups growing together in a primary throat culture* The more conspicuous colonies are the less important; they are the larger conical colonies surrounded by a wide zone of haemolysis. They belong to group C. The more numerous, smaller, slightly haemolytic ones belong to group A. (*Blood agar A, 18 hours at 37°C; reflected + transmitted light, x6.*)

12 *'Curled-leaf' colonies of a streptococcus* Some colonies dry out and curl up at the edges after a few days growth. This occurs commonly in streptococci of groups C and G, and less often in groups A and O, and indeed in other species. The group G streptococcal colonies shown here have this 'curled-leaf' appearance after overnight incubation. Their conical centre is not a constant feature of 'curled-leaf' colonies. The more numerous small, non-haemolytic colonies are typical nasal diphtheroids; note their regular dome-shape and fine matt surface. (*Primary culture from a nasal swab on blood agar A, 18 hours at 37°C; reflected + transmitted light, x6.*)

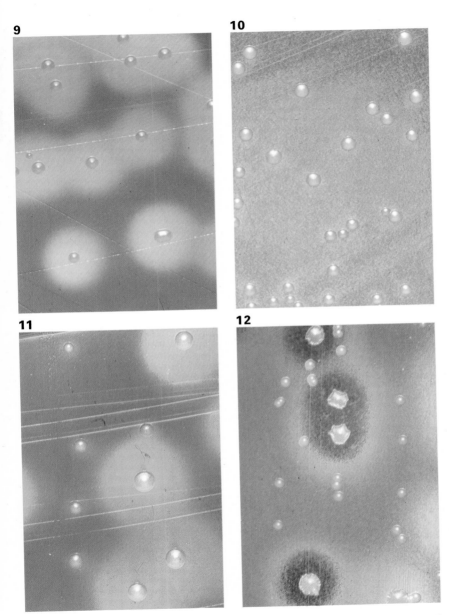

17

13 *Colonies of a haemolytic group B streptococcus* The colonies are rather large for streptococci, and they are smooth, convex and quite translucent, except for a slightly more opaque centre. Some group D streptococcal colonies have the same appearance. (*Blood agar A, 18 hours at 37°C; reflected + transmitted light, x6.*)

14 *'Draughtsman' colonies of pneumococcus* Typical colonies of the pneumococcus are flat with raised margins. Sometimes multiple concentric ridges produce a colony which is frequently likened to a 'draughtsman'. (*Blood agar A, 18 hours at 37°C; reflected light, x6.*)

15 *Another group of pneumococcus colonies* These are viewed by transmitted light to show alpha-haemolysis. Also included are a haemolytic staphylococcus and two non-haemolytic diphtheroids. (*Blood agar A, 18 hours at 37°C; transmitted light, x6.*)

16 *Mucoid colonies of the pneumococcus* At 18 hours the more isolated colonies are still large, moist and convex. Others have collapsed in the centre. (*Blood agar A, 18 hours at 37°C; reflected light, x6.*)

13

14

15

16

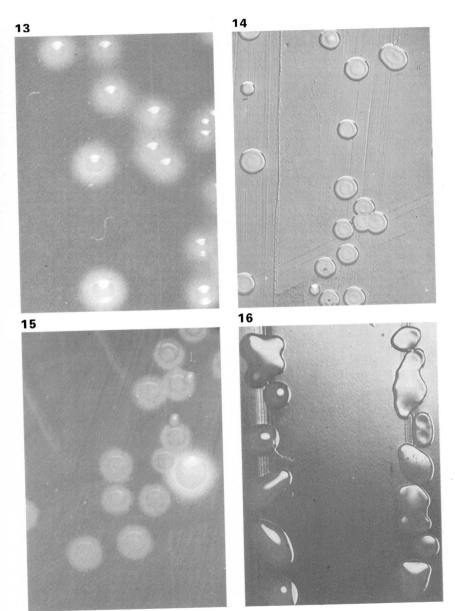

17 *Virtually pure culture of pneumococci from a sputum sample*
The same specimen was sown on blood agar and on heated blood agar ;
greening is more obvious on the latter. Since the red cells are broken by
heating blood agar, it is obvious that the intact red cell is not needed
for the demonstration of alpha haemolysis, as it is for beta haemolysis.
In alpha haemolysis the main effect is on the haemoglobin ; in beta
haemolysis the main effect is on the red cell membrane. (*Blood agar B
and heated blood agar B, 24 hours at 37°C; reflected light, x1.*)

18 *Colonies of Neisseria gonorrhoeae* After incubation for 24
hours in carbon dioxide, colonies of the gonococcus are small, discrete,
transparent and hemispherical. (*Heated blood agar B, 24 hours in CO_2
at 37°C; reflected light, x6.*)

19 *Colonies of Neisseria meningitidis* At 24 hours the colonies
are circular, greyish, transparent, lenticular discs. Their surface is smooth
and glistening. Contiguous colonies may merge to form a mucoid streak
of growth. They are oxidase positive. Older colonies are less distinctive.
(*Heated blood agar, 24 hours in 10% CO_2; reflected light, x6.*)

20 *The use of the oxidase test in finding neisserias* A solution
of oxidase reagent was poured on a plate on which a mixed culture had
grown from throat and nose swabs. Within a few seconds many of the
colonies had turned deep purple ; these were neisserias. Since growth
can be obtained from a reacting colony if it is subcultured promptly, this
is a useful method for selecting neisserias from the other colonies among
which they are likely to be found in cultures from the respiratory or
genital tracts. (*Blood agar A, 18 hours at 37°C, treated with 1% tetra-
methyl-para-phenylenediamine solution; reflected + transmitted light,
x1.*)

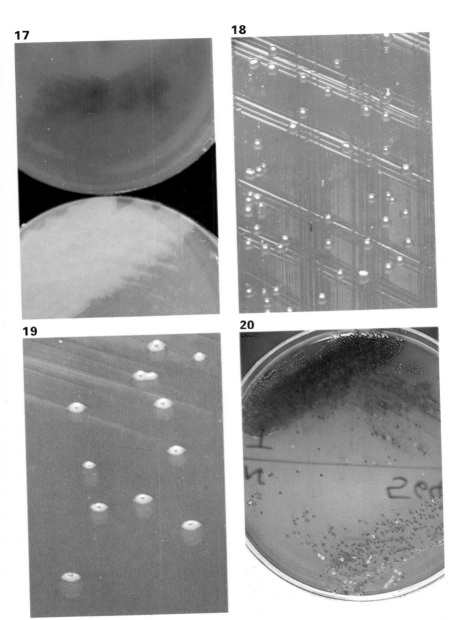

17

18

19

20

21

21 *Colonies of Neisseria sp. on blood agar* Colonies grown from a throat culture are viewed through the bottom of a petri dish. The colonies which concentrate the light through the medium are neisserias. When it is observed, this lens effect is useful for the presumptive identification of neisserias. (*Blood agar A, 18 hours at 37°C; transmitted light, x6.*)

22 *Colonies of Neisseria mucosa on MacConkey agar* Part of the large, mucoid, non-lactose-fermenting colony has remained attached to the wire. This isolate gave a prompt oxidase reaction (not shown). In some strains the mucus may be sufficiently dense to mask the oxidase reaction; this difficulty may be overcome by using washed cells in the test. (*MacConkey agar B, 48 hours at 37°C; reflected light, x6.*)

23 & 24 *Colonies of Corynebacterium haemolyticum* At 24 hours (**23**) the colonies are small, greyish, have a finely matt surface and are relatively undifferentiated. There is no obvious haemolysis. (*Reflected + transmitted light.*)

At 48 hours (**24**) the colonies have developed a zone of haemolysis, and they have a dark central spot. If the colony is lightly scraped aside (*arrow*) it is found that the dark central spot is left behind and that the colony has etched into the medium. (*48 hours at 37°C; transmitted light.*) (*Blood agar A, x6.*)

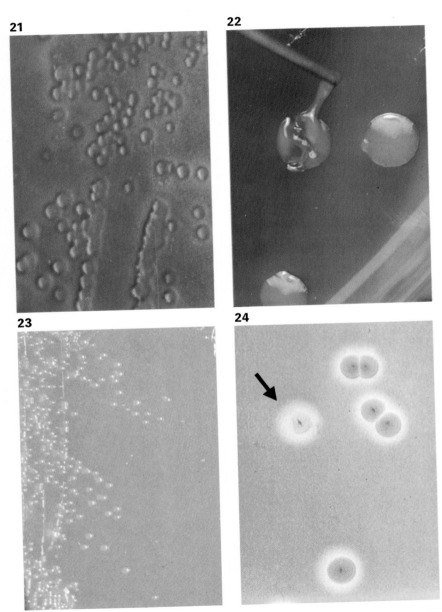

25, 26 & 27 *Colonies of Corynebacterium diphtheriae* On McLeod's tellurite medium after 48 hours incubation one sees the characteristic colonies by which the biological type of *Corynebacterium diphtheriae* is confirmed. (*McLeod's tellurite blood agar, 48 hours at 37°C; reflected light, x6.*)

Colonies of the *gravis* type (**25**) are flattened cones with frosted, metallic grey centres, and radially striated margins. These are typical 'daisy head' colonies on McLeod's medium.

Colonies of the *mitis* type (**26**) have a black raised centre which slopes gently down to a pale rim. This is the 'coolie-hat' colony which is common among *mitis* strains.

Colonies of the *intermedius* type (**27**) are much smaller than those of *gravis* or *mitis.* Typical colonies, with their dark centres and lighter margins, have been likened to frog's eggs.

28 *Corynebacterium ulcerans* This grows well on McLeod's tellurite medium. Some strains may be even larger than this, and may develop the radial striations typical of gravis type diphtheria bacilli. The lesion it produces in the skin of the guinea-pig (**379**) provides a means of recognition of *C. ulcerans.* It has caused tonsillitis in man and mastitis in cattle. (*McLeod's tellurite 'chocolate' agar, 48 hours at 37°C; reflected light, x6.*)

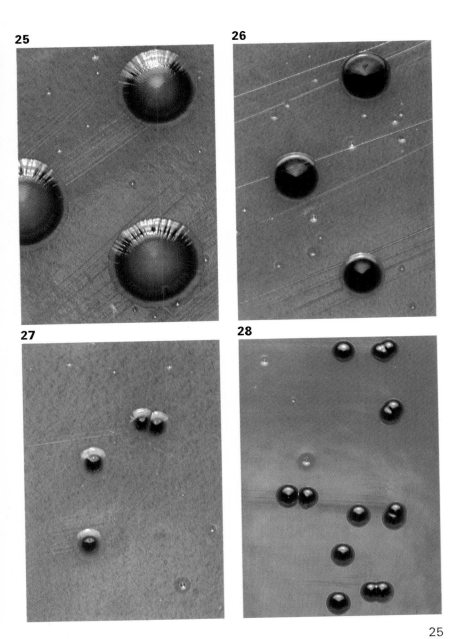

25

29, 30 & 31 *Colonies of Corynebacterium diphtheriae types on blood agar* Although the differences between the types are less obvious on blood agar than on McLeod's medium, they are still sufficiently characteristic for a diagnosis to be made at 18 hours. (*Blood agar B, 18 hours at 37°C; x6.*)

Gravis colonies (**29**) are large and have a dull matt surface and faint radial striations at the periphery. Some *gravis* strains are haemolytic, but the one shown here is not. (*Reflected light.*)

Mitis colonies (**30**) have an almost smooth surface, and they are surrounded by a narrow zone of haemolysis. (*Reflected + transmitted light.*)

Intermedius colonies (**31**) are non-haemolytic and are smaller than those of the other types. (*Reflected light.*)

32 *Two members of the normal nasal flora* The numerous small colonies are diphtheroids. They are greyish-white, dry, rough, and are flat cones. The other larger colonies with the moist, smoother surface are non-pathogenic neisserias. (*Blood agar B, 18 hours at 37°C; reflected light, x6.*)

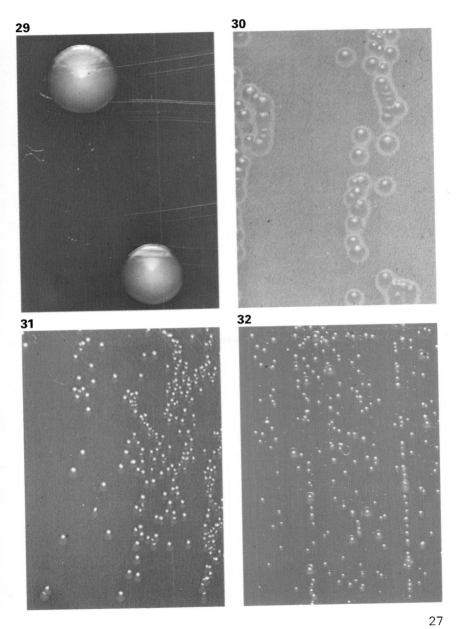

29

30

31

32

33, 34, 35, 36 & 37 *Use of growth temperatures in identification of mycobacteria* This is the first step in the scheme of Marks (1972) for the identification of mycobacteria encountered in Britain. A heavy suspension free of clumps is sown on four Löwenstein-Jensen slopes. One sample is incubated at each of the following temperatures : 25°, 37°, 42° and 45°C. After an appropriate period of incubation, the growths on the four slopes are compared. (*Reflected light, x1.*)

BCG (**33**) is a strict mesophile. It grows only at 37°C. Other strict mesophiles include *M. tuberculosis* and *M. bovis.*

M. marinum (**34**) is a psychrophile. It grows best at 25°C, and less well at 37°C.

M. gordonae (**35**) is a mesophile. It grows best at 37°C, but also grows at 25°C. Some strains grow at 42°C. The mesophile group includes also *M. flavescens, M. kansasii, M. fortuitum, M. terrae,* and some of the *M. avium-intracellulare* group.

M. smegmatis (**36**) grows over a wide range of temperatures. This group includes also *M. phlei* and some of the *M. avium-intracellulare* group.

M. xenopi (**37**) is a thermophile. The strain shown grew at 42°C and 45°C, but others grow also at 37°C.

33

34

35

36

37

38 *Growth of Mycobacterium tuberculosis on Löwenstein-Jensen medium* In the example shown the colonies are rough and flat with a raised centre and are buff coloured. The growth is friable and difficult to remove from the medium. Younger colonies, or colonies on poorer medium would be paler. Sometimes the colonies are more verrucose than those shown. (*Löwenstein-Jensen medium, 6 weeks at 37°C; reflected light, x1.*)

39 *Comparison of the growth of Mycobacterium bovis on Löwenstein-Jensen medium with glycerol, left, and with pyruvate, right* Strains of bovine tubercle bacilli grow very much better on the pyruvate-containing medium. Some strains of other mycobacteria from patients undergoing chemotherapy also prefer media containing pyruvate. (*Löwenstein-Jensen medium, 6 weeks at 37°C; reflected light, x1.*)

40 & 41 *Action of light on mycobacteria* Two slopes were sown with the same culture. The one on the left was exposed to light during growth; that on the right was grown in the same incubator but was shielded from light.

M. kansasii (**40**) produces pigment only in the light. It is therefore called a photochromogen. The photochromogens include some important pathogens.

M. gordonae (**41**) produces pigment in both light and dark. It is therefore called a scotochromogen. Most of the scotochromogens are non-pathogenic.

(*Löwenstein-Jensen medium, 6 weeks at 37°C; reflected light, x1.*)

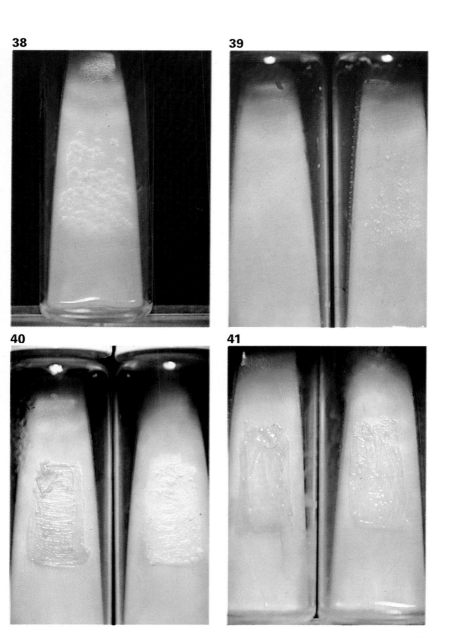

42 Growth of Mycobacterium intracellularis on Löwenstein-Jensen medium The growth is very light and is described as dysgonic. Compare it with the eugonic growth of *M. gordonae* (**41**), and of *M. kansasii* (**40**). (*Löwenstein-Jensen medium, 6 weeks at 37°C; reflected light, x1.*)

43 Growth of Mycobacterium avium on Löwenstein-Jensen medium Growth is even poorer than that of *M. intracellularis,* and can scarcely be seen. (*Löwenstein-Jensen medium, 6 weeks at 37°C; reflected light, x1.*)

44 Growth of mammalian tubercle bacilli in semi-solid medium *Mycobacterium tuberculosis,* left, is a strict aerobe and grows only near the surface of the medium. *M. bovis,* right, is a micro-aerophile, and grows well to a depth of about 1.5cm; it may have a band of densest growth in the depths of the medium as shown here. Aerobes are eugonic on solid media incubated aerobically; micro-aerophiles are dysgonic. (*Marks semi-solid medium, 3 weeks at 37°C; indirect transmitted illumination, x¾.*)

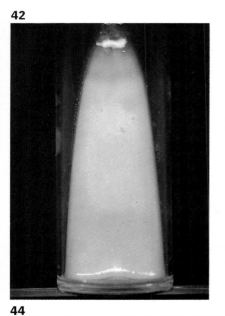

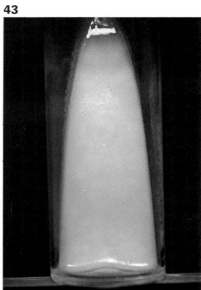

45, 46, 47 & 48 *Colonies of Bacillus spp.* These are usually large, flat, opaque and rough with an irregular edge. They are frequently hae-molytic. Some strains, for example those shown in **46** and **47**, are mucoid at first ; these later dry out to the more characteristic rough colony. (*Blood agar B, 18 hours at 37°C; reflected + transmitted light, x6.*)

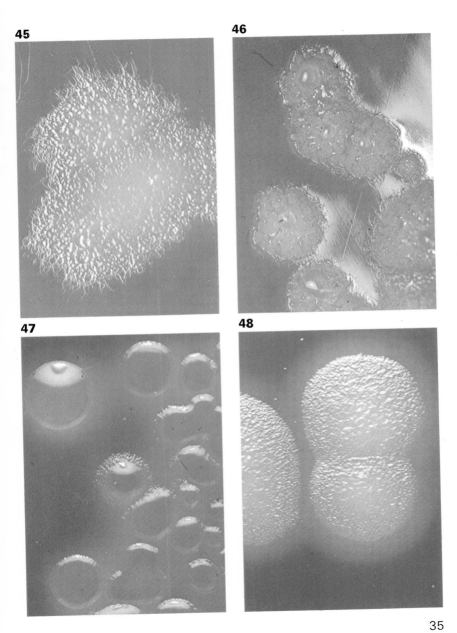

45

46

47

48

35

49 *Part of a Bacillus colony of mycoides type* This shows a characteristic feature of many colonies of *Bacillus spp.* – they grow over the surface of the medium by the extension of filamentous chains of cells. This produces a flat colony of interlaced cords with a rhizoid growing edge. (*Blood agar A, 18 hours at 37°C; reflected light, x6.*)

50 *Colonies of the anthrax bacillus* Colonies of *Bacillus anthracis* are large, opaque, white, and have a very rough surface and an irregular edge ; in these respects they are like the colonies of many other *Bacillus spp.* Characteristically, anthrax colonies are non-haemolytic and have the curled edge shown by some colonies in this figure. (*Blood agar B, 18 hours at 37°C; reflected light, x6.*)

51 *Bacillus anthracis colony* Under the low power of the microscope the edge of the colony of *Bacillus anthracis* has the appearance of curled hair-locks. (*Digest agar, 24 hours at 37°C; unstained, x28.*)

52 *Deep colonies of Bacillus anthracis* These have a filamentous appearance which has been likened to knotted string. This may be useful in the culture of the organism from contaminated specimens. Suspensions of wool or hair are heated to 70°C for ten minutes, then serial dilutions are mixed with agar which is poured into petri dishes. After 16 hours incubation the characteristic deep colonies may be sub-cultured for further study. (*Digest agar B, 18 hours at 37°C; indirect transmitted light, x6.*)

49

50

51

52

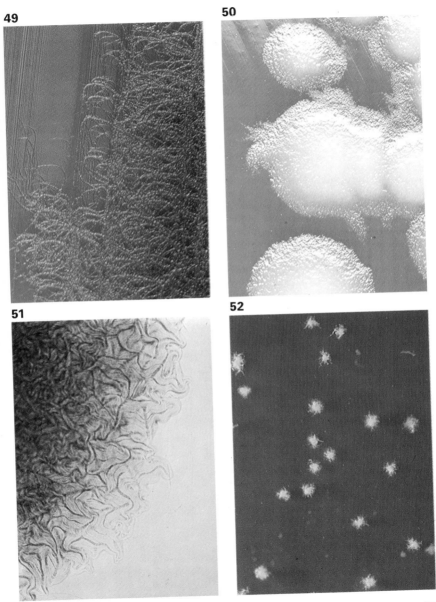

53 *Colonies of Clostridium welchii on blood agar* On blood agar they often have a target appearance ; the colony is surrounded by a clear zone, which is further surrounded by a zone of partial haemolysis. The inner zone is caused by the theta toxin and the outer one by the alpha. (*Blood agar B, 2 days anaerobically at 37°C; reflected + transmitted light, x6.*)

54 *Colonies of Clostridium welchii from a case of food poisoning* Many of the strains of *C. welchii* associated with food poisoning have been described as non-haemolytic, because they are not surrounded by the inner zone of complete haemolysis (cf. **53**). However they do produce lecithinase which causes the zone of partial clearing of the blood agar. (*Blood agar B, 18 hours anaerobically at 37°C; reflected + transmitted light, x6.*)

55 *Colonies of Clostridium sporogenes on blood agar* The raised glistening centre is surrounded by a mat of rhizoid filaments. The illumination is intended to show the surface of the colonies ; it does not show the haemolysis around each colony. (*Blood agar B, 2 days anaerobically at 37°C; reflected light, x6.*)

56 *Parts of two adjacent colonies of Clostridium tetani* Each consists of a flat network with a delicate edge of projecting filaments, by which the colonies rapidly spread over the surface of a blood-agar plate. Later, dense growth would show haemolysis. (*Blood agar B, 18 hours anaerobically at 37°C; reflected light, x6.*)

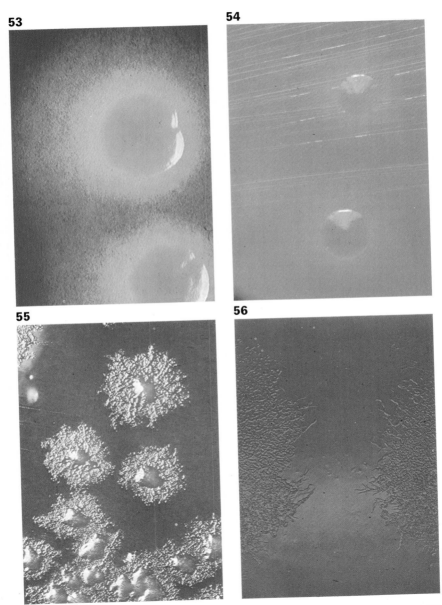

57 *Test for motility of clostridia in semi-solid media* Since clostridia lose their motility under aerobic conditions, some of the routine tests for motility are inapplicable. The test organism may be stabbed into a suitable semi-solid medium, and the pattern of growth examined. *C. welchii,* left, is confined to the line of the inoculum, because it is non-motile. *C. bifermentans,* right, has migrated through the medium, and is thus shown to be motile. Each species fails to grow in the aerobic zone at the top of the medium. (*Brewer's fluid thioglycolate medium, 24 hours at 37°C; indirect transmitted light, x¾.*)

58 *Action of clostridia on litmus milk* All three species have pro-duced acid, and all three have bleached the litmus in the depths of the medium. These are the only obvious changes with *C. tetani*, right. *C. welchii,* left, has produced a clot which has been disrupted by gas – the typical 'stormy clot'. With *C. sporogenes*, centre, the casein has been precipitated ; subsequently it would be digested by proteolytic enzymes. (*Litmus milk with iron strip, 2 days at 37°C; reflected light, x¾.*)

59 *Deep growth of clostridia in tubes of glucose agar* Each organism was sown into molten glucose agar which was then solidified and incubated. All three species have produced gas. This is most marked with *C. welchii*, left, and *C. tetani*, right, which are predominantly saccharolytic species. *C. sporogenes* is predominantly proteolytic, but it still produces a little gas. Each of these anaerobes fails to grow in a zone at the top of the medium in contact with the air. (*Glucose agar, 2 days at 37°C; reflected light, x¾.*)

60 *Cultures of, left to right, Clostridium welchii, C. sporogenes and C. tetani in cooked meat* The proteolytic *C. sporogenes* has turned the meat black, and has digested the sharp edges off the meat granules. (In the laboratory the most obvious feature of this species is its putrid odour.) All three species have produced much gas ; in two this has blown the plug of meat particles up the tube. (*Cooked meat medium, 24 hours at 37°C; reflected light, x½.*)

57

58

59

60

61 *Colonies of Actinomyces israelii on blood agar* The white, irregular, heaped colonies have a distinctive appearance like the crown of a molar tooth. They are adherent to the medium. (*Blood agar B, 5 days anaerobically at 37°C; reflected light, x6.*)

62 *Deep growth of Actinomyces israelii in shaken glucose agar* After incubation in air a band of growth is obvious about 2cm below the surface of the medium. Fewer colonies are growing in the anaerobic part of the medium, but there are none immediately next to the air. (*Glucose agar shake culture, 5 days at 37°C; indirect transmitted light, x1.*)

63 & 64 *Colonies of two strains of Nocardia asteroides on blood agar* In each case the colonies are white, opaque and firmly adherent to the medium in which they form a slight depression.

The colonies in **63** are heaped into fluted columns. In plan they are star shaped, hence the name *asteroides*.

In the strain in **64** aerial hyphae give the surface of the colonies a cottony texture. In some strains, colonies of this and the previous type may be found together.

(*Blood agar B, 4 days at 37°C; reflected light, x6.*)

61

62

63

64

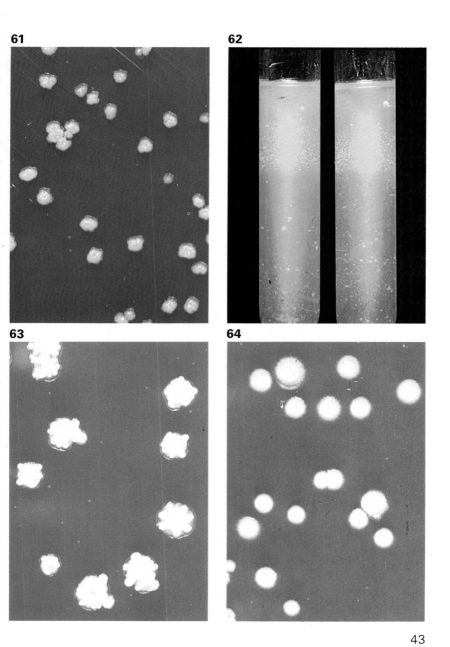

65 *Two strains of Nocardia madurae growing on Sabouraud's glucose agar* The growth is opaque and is irregularly heaped. In one strain it is creamy yellow and in the other it is developing the distinctive pink pigmentation. Many of the individual colonies are umbonate. The horny growth is quite adherent to the medium and is difficult to emulsify. (*Sabouraud glucose agar, 2 weeks at 37°C; reflected light, x1.*)

66 *Listeria monocytogenes in mixed culture viewed by indirect transmitted illumination* The colonies are translucent, bluish grey and are up to 1.5mm in diameter at 24 hours. If examined by reflected light they would show a finely textured surface and a watery consistency. The larger, opaque, creamy yellow colonies are *Staph. aureus*. Gray & Killinger (1966) recommend this method for detecting listeria in mixed culture and they give a detailed description of the method of illumination. (*Digest agar B, 24 hours at 37°C; indirect transmitted illumination, x6.*)

67 & 68 *Colonies of lactobacilli from the vagina* Lactobacilli grow well on heated blood agar plates incubated in an atmosphere containing CO_2. The colony morphology depends not only on the species but also on environmental conditions such as the dryness of the medium.

Lactobacillus colonies, as in **67**, are usually very small but they may be numerous. They have a rather rough surface and an irregular edge.

In some strains (**68**) the colonies are larger, flatter and have a very rough surface. Such colonies often remain intact when an attempt is made to prepare a smear from them.

(*Heated blood agar A, 24 hours in 5% CO_2 at 37°C; reflected light, x6.*)

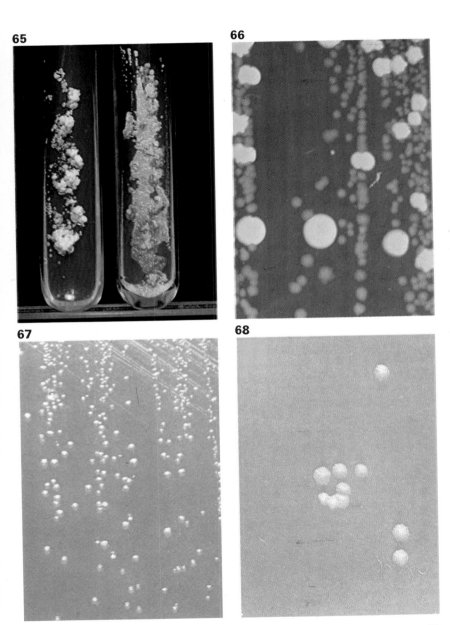

65

66

67

68

69 *Coliform colonies on digest agar* Most members of the Enterobacteriaceae (enteric gram-negative rods which are facultative
anaerobes) produce this type of colony on digest agar. They are about
3mm in diameter, circular, smooth, entire edged, low-convex, translucent,
greyish, butyrous and readily emulsified. Rougher forms are fairly
common. To assist colony identification, faecal specimens are usually
cultivated on a selective medium which incorporates an indicator
system, such as MacConkey, Leifson or Wilson and Blair medium.
(*Digest agar B, 18 hours at 37°C; reflected light, x6.*)

70 *Red colonies of the common lactose-fermenters of faeces*
The larger colonies of *Escherichia coli* are readily distinguished from
the smaller enterococci – a convenient name for *Streptococcus faecalis*
and related species which occur in the gut. Although they are fairly well
spread, it would be difficult to pick one of these *E. coli* colonies without
contamination by the streptococcus. One should always subculture
from an inhibitory medium to a non-inhibitory one to obtain a single-
colony strain for pure culture study. (*MacConkey agar B, 24 hours at
37°C; reflected light, x6.*)

71 *Colonies of enteric bacteria on MacConkey agar* The large,
pink colonies are the lactose-fermenters, *E. coli*. The creamy white
colonies are *Salmonella sp*. There is also a small colony of lactose-
fermenting enterococci. (*MacConkey agar B, 18 hours at 37°C; reflected
+ indirect transmitted light, x6.*)

72 *Colonies of Staphylococcus aureus, Pseudomonas
aeruginosa and Escherichia coli in mixed culture* Can you
allocate each colony to one of these species ? (*Digest agar B, 18 hours
at 37°C; reflected + transmitted light, x6.*)

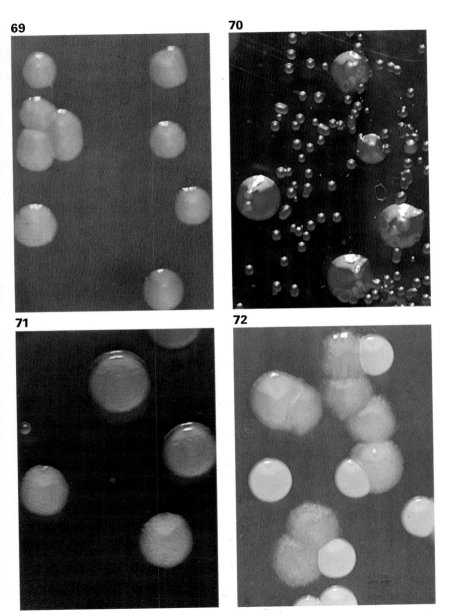

73, 74 & 75 *Klebsiella pneumoniae growing from a sputum sample* *(18 hours at 37°C, reflected light.)* A view of the whole plate (**73**) shows virtually pure growth of *K. pneumoniae*. (*Blood agar A, x½.*) The colonies (**74**) are very large, raised and viscid. (*Blood agar A, x6.*) The colonies of this strain on MacConkey agar (**75**) are even larger than on blood agar. This results from the greater supply of lactose in the MacConkey agar from which abundant extracellular polysaccharide is formed. (*MacConkey agar B, x8.*)

76 *Colonies of Klebsiella aerogenes on MacConkey agar* This species ferments lactose but does not greatly reduce the pH on MacConkey agar. Moreover, a reversion of pH quickly occurs in the area of confluent growth. Isolated colonies are large, domed, smooth and viscid. Compare the pale pink of these isolated colonies with the deep red of *Escherichia coli* colonies (**70**) on the same medium. (*MacConkey agar B, 18 hours at 37°C; reflected light, x½.*)

73

74

75

76

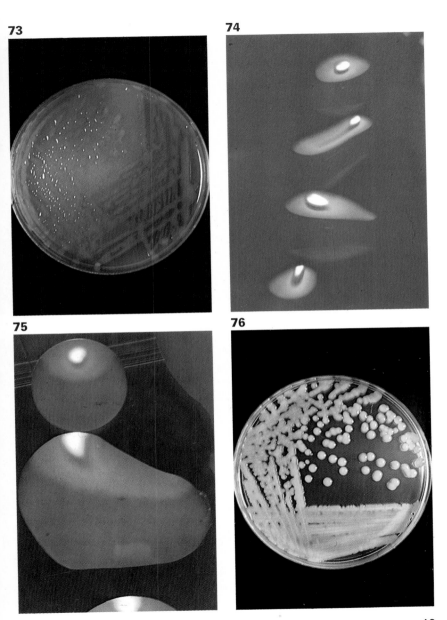

49

77 Swarming of Proteus mirabilis P. mirabilis and P. vulgaris (which, despite its name, is less common than P. mirabilis) both swarm over the surface of the common laboratory medium. Most characteristically, this produces a series of contour lines around a point of inoculation, which results from alternate periods of growth and of swarming, as shown here. These zones are less obvious on moist media, over which the organisms swarm without stopping. (*Digest agar B, 18 hours at 37°C; reflected light, x½.*)

78 Dienes test for identity of Proteus strains Identical strains of swarming proteus merge; others remain separated by a narrow line. Of the four strains spotted near the edge of this blood agar plate, only one is homologous with the strain in the centre. (*Blood agar A, 24 hours at 37°C; reflected light, x½.*)

79 & 80 'Sandwich plate' for the control of swarming of proteus
Because *Proteus spp.* are troublesome in the isolation of pure cultures of other bacteria, many methods have been devised to inhibit their swarming. In most methods an inhibitor is incorporated in the medium, e.g. the bile salts in media for growth of enteric bacteria. The advantage of the method illustrated here is that no special medium is required. The sample is streaked in the usual way on the surface of a blood agar plate. On the sown plate a layer of digest agar is poured. When this has set, its surface is flooded with spirit. Then it is thoroughly dried in an incubator, and finally incubated normally. (*18 hours at 37°C; x½.*)

Figure **79** shows a plate so treated. The bacterial growth is sandwiched between the blood agar and digest agar layers; here proteus does not swarm. Colonies can be seen through the surface and picked with a wire for further study. The colonies with a broad band of haemolysis are *Streptococcus pyogenes*. (*Blood agar B + digest agar B; reflected + transmitted light.*)

Figure **80** is a blood agar plate on which the same specimen was sown, but with no additional layer of agar. Proteus has swarmed over the surface and obscured the causative streptococcus. (*Blood agar B; reflected light.*)

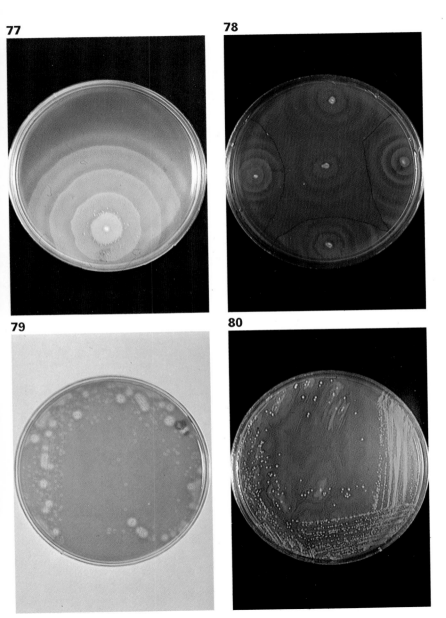

77

78

79

80

81 & 82 *The relative selectivity of MacConkey and Leifson media* The same faecal specimen was sown on to these two plates. Only lactose-fermenters are evident on the MacConkey plate (**81**). Most of the isolated colonies on the Leifson plate (**82**) are salmonellas, but a few lactose-fermenters have grown ; in the more confluent areas of growth the salmonella colonies have black centres. Some of the isolated salmonella colonies have been removed for slide agglutination. (*18 hours at 37°C; reflected light, x½.*)

83 *Colonies of Escherichia coli on MacConkey medium* This is a hand lens view of **81**. The colonies are rose pink, and they are rougher and more opaque than those shown in **70**. This is an effect of the medium ; colonies usually appear rougher on our MacConkey A than on our MacConkey B medium. The opacity results from the precipitation of bile salts under the growing colonies. A few small enterococcal colonies are also present. (*MacConkey agar A, 18 hours at 37°C; reflected light, x6.*)

84 *Salmonella colonies on Leifson's medium* These non-lactose-fermenting colonies have a beaten copper surface and a slightly irregular edge. (Most salmonella strains on first isolation are smoother than this, some are rougher). In the area of denser growth colonies on the same plate had black centres. (*Leifson's deoxycholate citrate agar, 18 hours at 37°C; reflected light, x6.*)

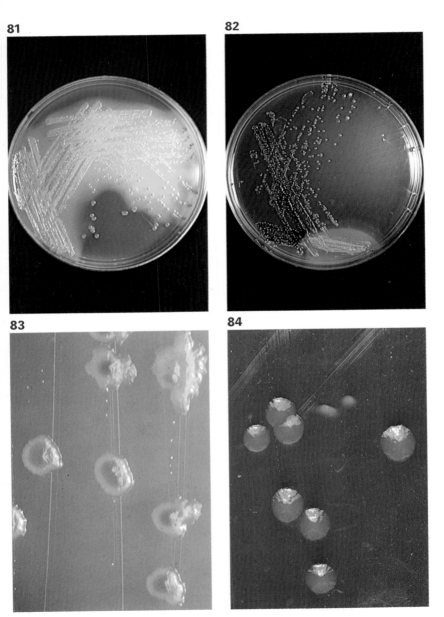

85 & 86 *Salmonella colonies on Wilson and Blair medium* The colonies of *S. typhi* in **85** are conical and smooth in the more heavily sown area of the plate, top, and have a central black area surrounded by a wide clear peripheral zone. More isolated colonies, bottom, have a larger black zone and a thinner clear margin ; their edge is slightly crenated and their tops are flatter, and the medium surrounding them contains dark precipitate.

The strain of *S. paratyphi B* (**86**) grows first as mucoid colonies which eventually flatten. The sequence is illustrated from top to bottom in this photograph. The colonies are becoming progressively more collapsed until the one at the bottom has a crater with a central knob. (*Wilson and Blair bismuth sulphite medium, 18 hours at 37°C; reflected light, x6.*)

87 *Colonies of Salmonella typhimurium viewed by indirect transmitted illumination* This type of lighting is particularly useful for scanning primary cultures on Leifson's medium for salmonellas, which have the distinctive fine regular granularity shown by these colonies (cf. **92**). (*Leifson's deoxycholate citrate agar, 18 hours at 37°C; indirect transmitted light, x6.*)

88 *Colonies of Salmonella arizona* The example shown is a prompt lactose-fermenter. It might therefore have been rejected as a non-pathogen, had it not been for its good growth on Leifson medium and its salmonella-like colonies on Wilson and Blair medium. (*Leifson's deoxycholate citrate agar, 18 hours at 37°C; reflected light, x6.*)

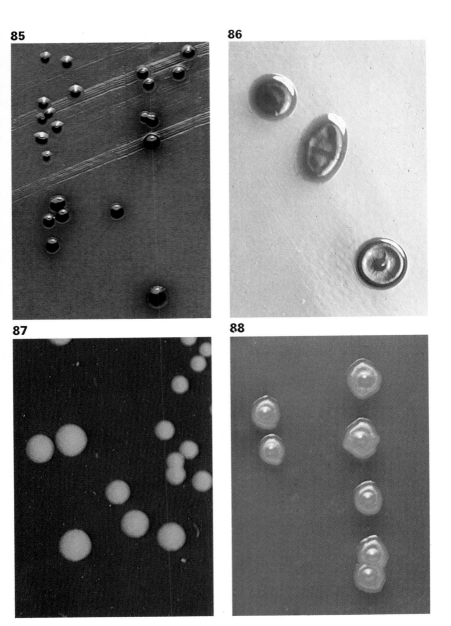

85

86

87

88

89 & 90 *Shigella sonnei colonies on MacConkey and on Leifson's medium* The two plates were sown with the same faecal specimen. On Leifson's deoxycholate citrate agar (**89**) the three pink colonies are lactose fermenters (coliforms). The others are *S. sonnei*; their colonies are fairly smooth and conical ('coolie hat' shaped), and have slightly pink centres and entire edges.

On MacConkey A agar (**90**) the *sonnei* colonies are rough and flat, and have an irregular edge, which in places is effuse and spreading. Fewer *sonnei* colonies grow on the MacConkey plate — compare the ratio of lactose fermenters to nonfermenters on these two media. Note also the very heavy growth of faecal streptococci on this medium. (*24 hours at 37°C; reflected light, x6.*)

91 *Colonies of Pseudomonas aeruginosa on deoxycholate citrate agar* The tenacious mucoid colonies of this non-lactose-fermenter have been likened to chewing gum. Note the consistency of the colony on the right which has been manipulated with a loop. (*Leifson's deoxycholate citrate agar, 18 hours at 37°C; reflected light, x6.*)

92 *Colonies of Pseudomonas aeruginosa viewed by indirect transmitted illumination* The irregular, coarse granularity of some of these colonies readily distinguishes them from salmonellas (cf. **87**). Moreover, the colonies are not as regularly circular as are those of salmonellas. (*Leifson's deoxycholate citrate agar, 18 hours at 37°C; indirect transmitted light, x6.*)

89

90

91

92

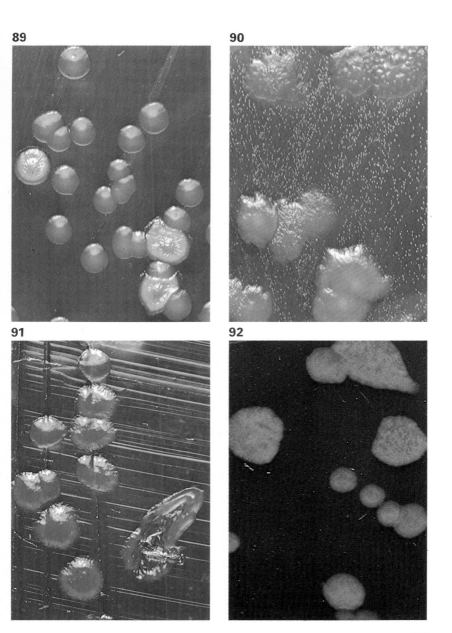

93 *Pseudomonas aeruginosa colonies on blood agar* Irregular colonies with a 'beaten-copper' surface and an effuse edge. This plate was sown with a specimen of sputum. (*Blood agar A, 24 hours at 37°C; reflected light, x6.*)

94 *Areas of confluent growth of Pseudomonas aeruginosa on a digest agar plate* Note the deep green colour of the growth. Plaques are frequently found in freshly isolated strains, especially in the area of confluent growth. In some strains these are caused by phage, in others by bacteriocins. (*Digest agar, 24 hours at 37°C; reflected light, x6.*)

95 *Three tubes of Pseudomonas aeruginosa in digest broth* The slight greening at the top of the left one results from oxidation of pyocyanin. Shaking the tube (middle) oxidises the pyocyanin throughout. The right hand tube was shaken with 1ml of chloroform. This extracts the blue pigment into the chloroform layer. This result is specific for *P. aeruginosa*. (*Digest broth, 24 hours at 37°C; reflected + transmitted light, x¾.*)

96 *Primary culture of sputum from a case of cystic fibrosis* The predominant bacterium is a mucoid *P. aeruginosa.* This is a common finding in this disease, but the reason for the association is not known. (*Blood agar A, 18 hours at 37°C; reflected light, x¾.*)

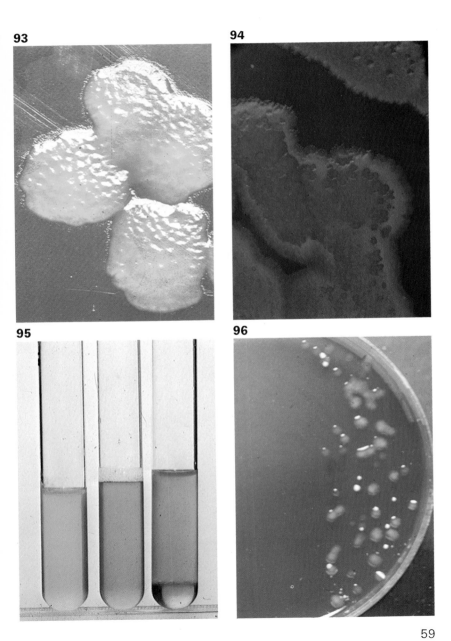

93

94

95

96

59

97 *Variability of pigment production by strains of Pseudo-monas aeruginosa* The characteristic green colour of cultures of *P. aeruginosa* results from its production of two pigments, the yellow fluorescein and the blue pyocyanin. Different strains produce these pigments in different proportions, as shown here. The tube on the left is unsown control medium and the one next to it contains a culture which is producing virtually no detectable pyocyanin. The next strain is producing abundant pyocyanin which makes the colour deep green, and the one on the right is producing a little. Special media are available for enhancing pigment production. (*Sabouraud maltose agar, 3 days at 37°C; reflected + transmitted light, x¾.*)

98 *Growth of some Pseudomonas spp. on potato* The growth of *P. stutzeri* (left) is a pale honey colour; *P. mallei* produces a café-au-lait growth. *P. pseudomallei* (right) has produced a thin whitish growth which is difficult to see; this organism has characteristically discoloured the potato plug. (*Potato plugs incubated in a humid atmosphere, 5 days at 37°C; reflected light, x¾.*)

97

98

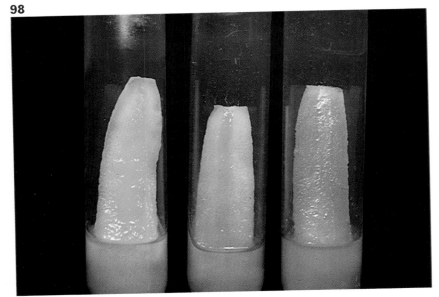

99 *Pseudomonas stutzeri colonies* This species gives rise to a variety of colony types. Most of these colonies are smooth, but varying grades of roughness are also shown. The roughest colonies are large and flat; subcultures of these yield rough colonies only. Subcultures of smooth colonies yield a variety of colony types. (*Digest agar B, 24 hours at 37°C; reflected + indirect transmitted light, x6.*)

100 *Vibrio cholerae colonies on thiosulphate, citrate, bile-salts, sucrose agar* The colonies are yellow, translucent, flat discs with entire edges. This appearance is useful in selecting the organism in mixed culture from faeces. (*Difco TCBS agar, 24 hours at 37°C; reflected light, x6.*)

101 *Vibrio cholerae on blood agar* Although it does not produce a soluble haemolysin, colonies of *V. cholerae* may be surrounded by a zone of clearing. Later this extends to clear the entire plate. This has been called haemodigestion. (*Blood agar B, 20 hours at 37°C; reflected + transmitted light, x6.*)

102 *Vibrio eltor on blood agar* El Tor strains haemolyse blood agar and produce a soluble haemolysin (**346**). The strain shown had a double zone of haemolysis : the outer zone was partially cleared at 18 hours and gradually became clearer — it appeared to correspond to the zone of haemodigestion of *V. cholerae;* the inner zone was clear throughout and presumably resulted from the action of the haemolysin. (*Blood agar B, 24 hours at 37°C; reflected + transmitted light, x6.*)

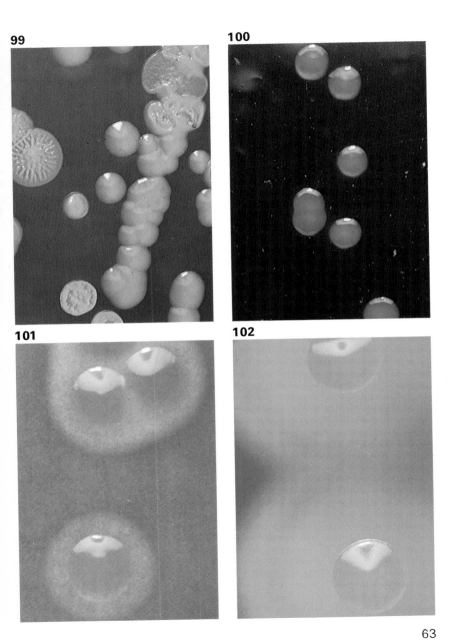

103 *Colonies of Yersinia pseudotuberculosis on digest agar*
The colonies have a granular surface, a raised, more opaque centre, and
a transparent effuse edge. (*Digest agar B, 24 hours at 37°C; reflected +
indirect transmitted light, x6.*)

104 *Yersinia pseudotuberculosis on blood agar at two days*
Colonies are differentiated less well than on digest agar, and even at
two days they do not show the effuse edge seen in **103**. At this stage
the blood agar plate had lysed and browned. (*Blood agar B, 2 days at
37°C; reflected light, x6.*)

105 & 106 *Cultures of Pasteurella septica on blood agar* The
colonies are water-clear and are at first raised; later they would flatten.
Each culture was grown from a cat bite wound. The colonies in **105** have
a domed centre and a sloping edge. They are oxidase positive. Figure
106 shows a mucoid strain with large, water-clear colonies. (*Blood agar
A, 18 hours at 37°C; reflected + transmitted light, x6.*)

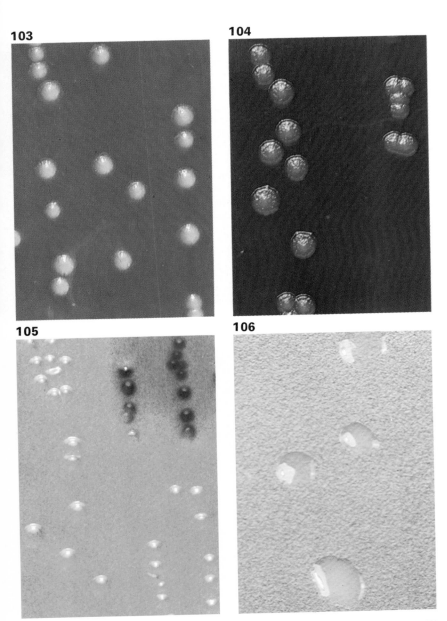

107 & 108 *Satellitism of other colonies by Haemophilus influenzae on blood agar* Blood agar has inadequate 'V' factor to permit full development of *H. influenzae* colonies. Some other colonies produce surplus 'V' factor which diffuses into the medium. This results in satellitism – the colonies of *H. influenzae* are large in the vicinity of a donor colony, and become progressively smaller the further they are from the donor. The staphylococcal colonies at the bottom of **107** are providing the 'V' factor. In **108** the haemophilus is satellitising pneumococcal colonies. (*Blood agar A, 18 hours at 37°C; reflected light, x6.*)

109 *Satellitism of staphylococcus colonies by Haemophilus influenzae* This whole blood agar plate was streaked to obtain isolated colonies and then spotted with *S. aureus* in three places. It gives a general view of how the colony size diminishes at a distance from the donor colony. (*Blood agar A, 24 hours at 37°C; reflected light, x¾.*)

110 *A comparison of the size of Haemophilus influenzae colonies on blood agar, left, and on heated blood agar, right* On heated blood agar the colonies are fully developed even in the absence of a donor organism, because the medium has adequate 'V' factor. The colonies are almost undetectable on the blood agar. (*Blood agar A and heated blood agar A, 18 hours at 37°C; reflected light, x6.*)

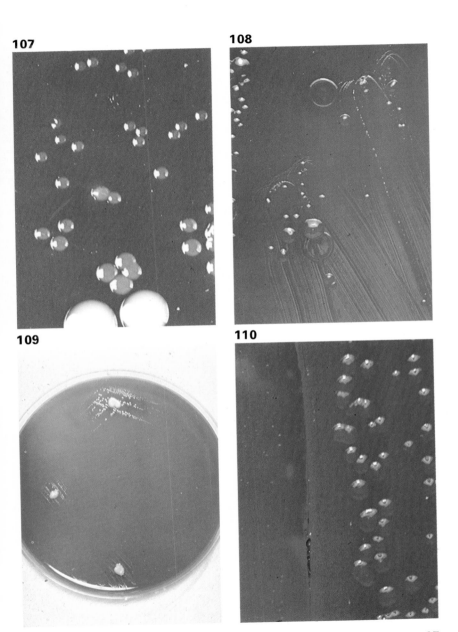

107

108

109

110

111 *Mucoid colonies of Haemophilus influenzae, type B, on Fildes' medium* Capsulated strains of *H. influenzae* produce large, mucoid colonies on Fildes' medium. With an appropriate method of illumination they would be iridescent, but this is not shown here. (*Fildes' medium, 24 hours at 37°C; reflected light, x6.*)

112 *Growth factor requirements of Haemophilus species* Two strains were sown on this peptone agar plate, one on each half. Discs containing X factor (brown), V factor (white) and both X and V (red) were then placed on the surface of the inoculated plate. Colonies of *H. influenzae*, left, grow in the vicinity of the disc containing X and V factors, but not near those containing either factor alone ; *H. influenzae* requires both factors. *H. parainfluenzae* requires V factor only ; this is demonstrated by its growth in the vicinity of the V factor disc as well as around the disc containing both factors. (*Peptone agar, 24 hours at 37°C; reflected light, x¾.*)

113 & 114 *Haemolytic colonies of Haemophilus species* These are often seen in throat cultures, and they may at first be confused with colonies of beta-haemolytic streptococci. Like neisserias, they often show a 'lens effect' (**21**) when viewed through the medium. This, and the two other features illustrated here, are helpful in their recognition. The more isolated they are (**113**) the smaller are haemophilus colonies and the smaller are their zones of haemolysis. Haemolytic haemophilus colonies have a characteristic translucent granularity (**114**) rather like fused crystals, and they are soft. Streptococcal colonies on the other hand are usually hard, white and opaque. (*Blood agar A, 18 hours at 37°C; reflected + transmitted light, x6.*)

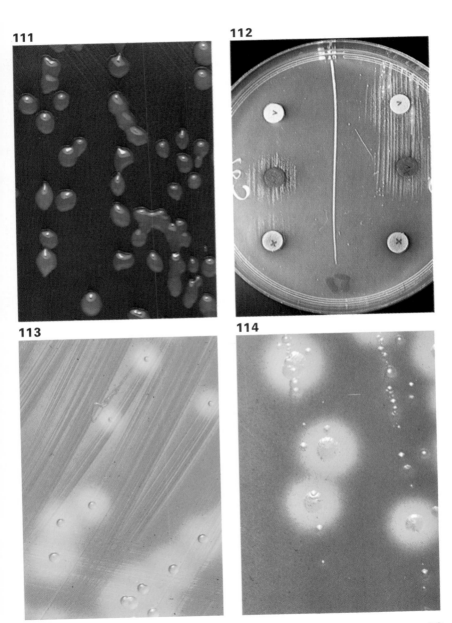

115 *Colonies of Actinobacillus aprophilus on blood agar*
Small, smooth, entire, slightly domed colonies, which may be adherent to the medium and produce pitting. With further incubation the colonies become rougher and more opaque. One of the small colony variants characteristic of this species is shown in the centre of the field. (*Blood agar B in 10% CO_2, 24 hours at 37°C; reflected light, x6.*)

116 *Colonies of Bordetella pertussis on Lacey's medium* The dome-shaped colonies have a smooth, metallic surface – note how clearly the lamp is imaged on each colony. This is a primary culture from a pernasal swab. (*Lacey's medium with antibiotics, 3 days at 37°C; reflected light, x6.*)

117 *Bordetella parapertussis growing on digest agar* By comparison with *B. pertussis, B. parapertussis* grows on relatively simple media. A characteristic brown pigment diffuses into the medium from areas of heavy growth. (*Digest agar B, 3 days at 37°C; transmitted light, x$\frac{3}{4}$.*)

118 *Bordetella bronchiseptica colonies on MacConkey agar*
B. pertussis and *B. parapertussis* will not grow on this medium. (*MacConkey agar B, 24 hours at 37°C; reflected light, x6.*)

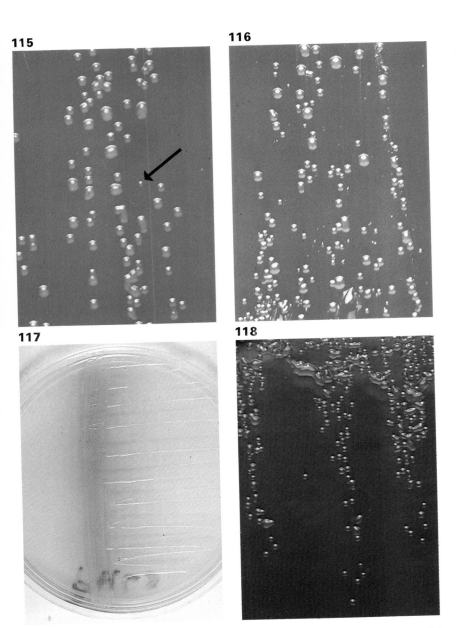

119 *Brucella abortus colonies on serum dextrose agar* At the appropriate angle of indirect transmitted illumination, the colonies are characteristically small and blue. This appearance is helpful in identifying *Brucella abortus* in mixed culture, especially if a comparison can be made of the growth in the presence and in the absence of carbon dioxide. (*Serum dextrose agar with antibiotics, 4 days at 37°C in CO_2; indirect transmitted light, x6.*)

120 & 121 *Dye-inhibition tests for the identification of Brucella species* Serum dextrose agar plates containing respectively 1 in 25,000 basic fuchsin and 1 in 50,000 thionin are used in the identification of brucellas. A loopful of each of three known strains has been streaked on quadrants of these plates. *B. abortus*, top right, grows only on fuchsin ; *B. suis*, top left, grows only on thionin ; *B. melitensis* grows on both. *B. suis* fails to grow in the presence of the basic fuchsin (**120**). *B. abortus* fails to grow on the thionin plate (**121**). (*5 days at 37°C in 10% CO_2; reflected light, x¾.*)

122 *Another method of testing dye sensitivity of a brucella* The organism was heavily sown on a blood agar plate in which two strips of dyed filter paper were submerged. This brucella has grown over the thionin-impregnated strip, left, but not over the basic fuchsin-treated strip, right. It is thereby identified as *B. suis*. (*Blood agar, 4 days at 37°C in 10% CO_2; reflected light, x¾.*)

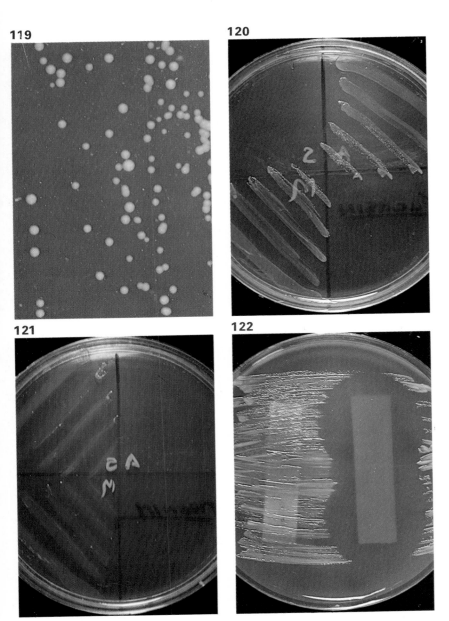

119

120

121

122

123 *Colonies of Streptobacillus moniliformis from the middle ear of a rat* The two larger bacterial-type colonies (right of centre) are surrounded by many smaller L-form colonies. An area on the left has been wiped with a loop to demonstrate that parts of the L-form colonies remain embedded in the medium. (*Sheep blood agar, 4 days at 37°C; reflected light, x6.*)

124 *Growth of Streptobacillus moniliformis in serum broth* (*left*) The tube on the right shows its failure to grow in digest broth. The tubes have been shaken lightly to disturb the deposited growth. The serum broth contains the characteristic 'snow-flakes' which do not disintegrate on shaking. (*30% horse serum digest broth, 48 hours at 37°C; indirect transmitted light, x¾.*)

125 *Colonies of Mycoplasma pneumoniae on Whittlestone's medium* These tiny colonies are viewed through the x4 objective of the microscope. They are rather variable in size, and they do not have the fried egg appearance common to many mycoplasma colonies. (*Whittlestone's medium in a moist atmosphere, 7 days at 37°C; transmitted light, x42.*)

126 *Mycoplasma pneumoniae growing in fluid medium* Growth of mycoplasmas in fluid media produces only faint turbidity. This may be difficult to appreciate because of inherent turbidity of the serum-rich medium itself. Acid produced from glucose is detected by the pH indicator included in the medium. Uninoculated control medium, left; medium growing *M. pneumoniae*, right. (*Whittlestone's medium, 5 days at 37°C; transmitted light, x¾.*)

123

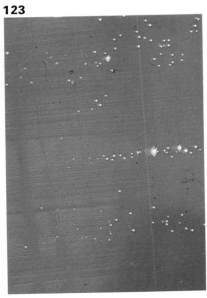

124

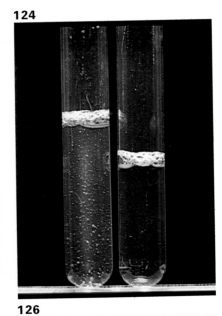

125

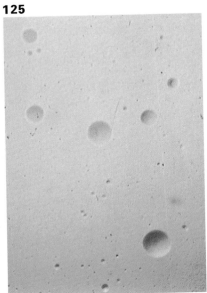

126

Bacteria: Microscopic Appearances

In other fields of biology, species are characterised by the morphological features of the individual. This applies to fungi too, but in bacteria (except when an organism is identified by staining with specific fluorescent antibody) the morphological features of the bacterial cell provide only a primary subdivision of this enormous group. A few morphological types can be recognized : cocci, rods, branching filaments and spirals.

Bacteria are divisible into two major groups according to their reaction to Gram's stain. The different staining results from the essentially different structure of the cell walls of gram-positive and gram-negative bacteria, and is correlated with many other important differences between the two groups.

Microscopy is most helpful in the study of the gram-positive rods ; further staining methods are often employed. Methylene blue and Albert's stains yield further information about corynebacteria. The Ziehl-Neelsen method distinguishes the mycobacteria. Nocardias are recognizable by their branching filaments. Microscopy is likewise very useful in the study of the filaments and clubs of *Actinomyces*, and the spores, capsules and motility of *Bacillus* and *Clostridium*.

The shape and arrangement of the cocci are helpful in their recognition.

The gram-negative rods constitute the commonest group encountered in most bacteriology laboratories. Unfortunately it is this group in which morphological study is of most limited use, but some features may be helpful. The extreme variation in size of *Proteus spp.* in different phases of growth is shown in **278** and **279**. Their diplobacillary arrangement is useful in *Moraxella*. The characteristic morphology of *Streptobacillus moniliformis* is well known, but similar forms may be seen in *Haemophilus spp.* and in other species with cell wall defects. The spirilla are helicoidal. Vibrios are curved bacteria having a comma shape ; their cultures often include spirillar forms. These spiral bacteria possess flagella and thereby differ from the spirochaetes which move by flexion of the cell body. Spirochaetes range in size from the large borrelias which stain readily by Giemsa, down to the extremely thin leptospiras which are viewed by dark-field illumination or by methods which greatly increase their diameter by plating them with silver.

The size of bacterial cells is not quoted in this section. Since virtually all the photomicrographs are reproduced at x1,000, the dimensions of a cell in μm can be obtained directly by measuring it with a millimetre rule.

Electron micrographs of the morphological components of bacterial cells are not included because they can be found in most standard textbooks of bacteriology and because they seem inappropriate in a colour atlas.

127 & 128 *Simple stains for bacteria* Bacterial cells are colourless, but with simple stains they can be seen readily under the microscope. Simple stains such as methylene blue, carbol fuchsin or nigrosin negative stain, are rapid and they show the shape of cells and how they are arranged. Because morphological features are of limited use in identifying bacteria, more complex differential stains are usually employed by bacteriologists.

Figure **127** is a simple negative stain of *Escherichia coli*; the nigrosin occupies the spaces between the cells. Rods are distributed singly throughout the field. (*Nigrosin, x1,000.*)

Figure **128** is a simple positive stain. The cells in this photograph are staphylococci and their shape and arrangement are consistent with this. (*Methylene blue, x1,000.*)

129 *A mixed smear stained by Gram's method* This mixture consists of the gram-positive *Staphylococcus albus* (purple-stained spheres arranged in grape-like clusters), and the gram-negative *Escherichia coli* (red-stained rods scattered singly throughout the field). The Gram stain is the bacteriologist's most important differential stain. The cell walls of gram-positive and gram-negative bacteria are essentially different, and this is correlated with other fundamental differences between organisms of the two groups. (*Gram, x1,000.*)

130 *A smear of micrococci from a contaminant colony* This smear was prepared from one of the colonies shown in **7**. It may be difficult to decide whether these organisms are arranged in pairs or tetrads. This is readily resolved by examining an over-decolorised area of the smear shown; in a number of groups one of four organisms stains differently from the other three. Gram-negative cells are found in older cultures of most gram-positive organisms. (*Gram, x1,000.*)

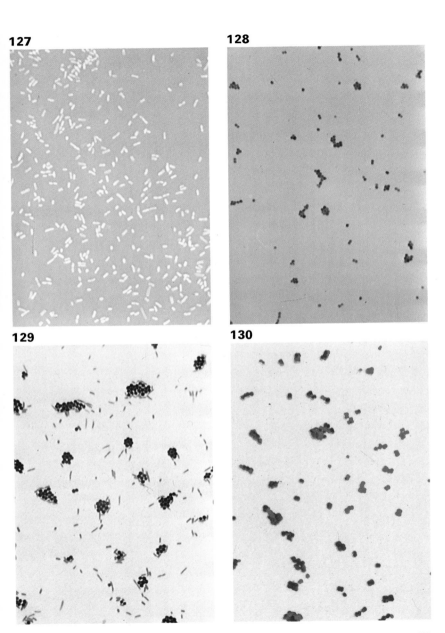

127

128

129

130

131 *Gram stain of pus from an abscess* Overlying the pus cell is a typical cluster of *Staphylococcus aureus*, looking like a small bunch of grapes. Paired and single cocci are also present. (*Gram, x1,000.*)

132 *Streptococci in a smear of pus* Long and short chains and a few pairs of gram-positive cocci are present. The pus cells are disintegrating. The smear was prepared from an empyema (pus in the thoracic cavity). The streptococcus cultivated from the pus was anaerobic. (*Gram, x1,000.*)

133 *Pneumococci in cerebrospinal fluid* There are many pairs of lanceolate diplococci. Most are gram-positive, but some, especially intracellular pairs, are gram-negative. This smear was prepared immediately after the CSF was taken from the patient ; the enormous number of bacteria present do not therefore represent multiplication in the sample after collection. (*Gram, x1,000.*)

134 *Smear from a culture of a pneumococcus* The morphology of the pneumococcus is most characteristic in the tissues, as shown in the previous figure. In culture, similar forms may be present, but often the cocci are arranged in chains, or they are more rounded, or sometimes more elongated and rod-like, as shown here. (*Gram, x1,000.*)

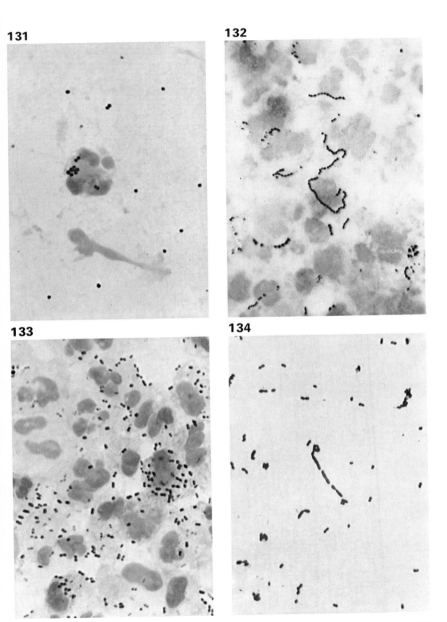

131

132

133

134

135 *Impression preparation of the peritoneum of a mouse infected with pneumococcus* The preparation has been stained to demonstrate the capsules. The bodies of the diplococci are stained red. These are enveloped by a blue-staining capsule. Red-stained nuclei of mouse tissue cells are grouped mainly in the top half of the photograph. (*Muir's capsule stain, x1,000.*)

136 *Gonococci in a smear of urethral discharge* There are numerous pus cells. Three of them in the middle of the field contain many gram-negative, oval diplococci whose apposed surfaces are flattened. Intracellular cocci of this appearance are typical of the gonococcus. (*Gram, x1,000.*)

137 *Smear from a culture of Neisseria gonorrhoeae* The appearance of the gonococcus in culture is quite different from that in tissue smears. Rounded cocci, variable in size and depth of staining are seen. The cells readily undergo autolysis. (*Gram, x1,000.*)

138 *Meningococci in cerebrospinal fluid* The gram-negative cocci are in pairs; some are intracellular. The long axes of pairs of cocci are parallel, and not in line as with the pneumococcus. (*Gram, x1,000.*)

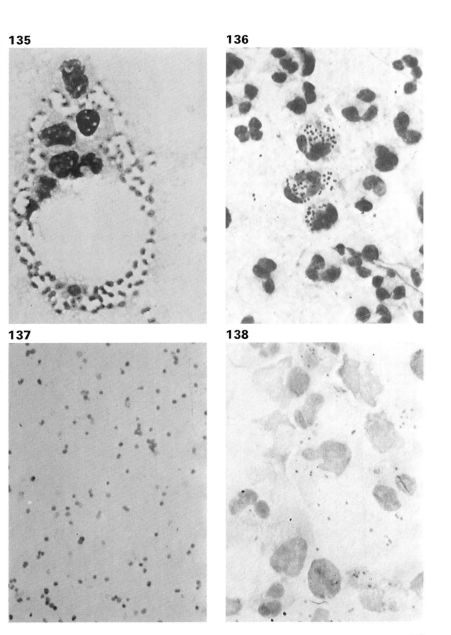

135

136

137

138

139 *Smear from a culture of a diphtheria bacillus* The club shape (thick at one end and tapering towards the other), the arrangement in so-called Chinese letter patterns, and the slight curvature of the rods are all typical of the corynebacteria. Gram-stained smears are of little value in the identification of *C. diphtheriae* ; methylene blue and special stains for metachromatic granules are more useful. (*Gram, x1,000.*)

140, 141 & 142 *Metachromatic granules of Corynebacterium diphtheriae types* Cultures on Loeffler's serum of all three types of *C. diphtheriae* contain metachromatic granules. The frequency of the granules and their location is useful in identifying the type of bacillus.

The *mitis* type appears in **140**. The bacilli are long, curved and thin, and some of them contain a number of granules. The bacilli stain pale grey-green, the granules bluish-black.

Diphtheriae bacilli of *intermedius* type (**141**) contain fewer metachromatic granules than do *mitis*. The cells are quite pleomorphic : some of them are long, others are quite short and many are club-shaped.

In the *gravis* type (**142**) metachromatic granules are very sparse, and in most fields no granules will be seen at all. The cells are short and stain quite evenly. (*Albert's stain, x1,000.*)

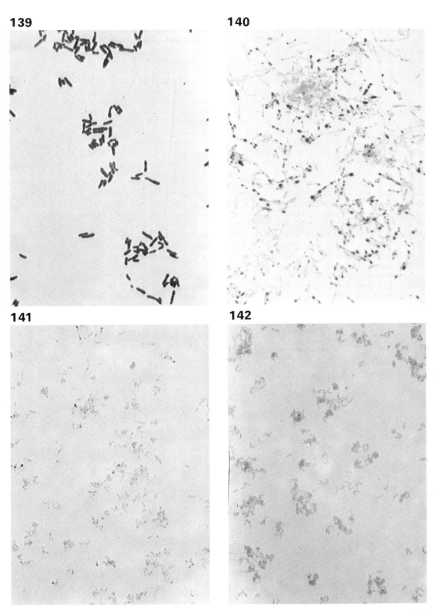

143 *Methylene blue-stained smear of Corynebacterium xerosis*
Many of the cells are quite long and all are distinctly barred. The cells
are arranged side by side (in palisades) or at angles to produce
'Chinese-letter-forms'. Club-shaped cells are rare in this species.
(*Loeffler's methylene blue, x1,000.*)

144 *Smear of tuberculous sputum* In this field there is a small
group of tubercle bacilli, a pair and a single tubercle bacillus. Yeasts
and streptococci are also present. This illustrates why treatment with
strong alkali or acid is used to decontaminate sputum in routine
laboratories. (*Ziehl-Neelsen, x1,000.*)

145 & 146 *Smears of deposit from centrifuged urine samples*
Urine from a case of tuberculosis (**145**) shows some single bacilli, but
others are arranged in groups which have remained attached, sometimes
side by side, after division. Individual cells stain unevenly and are fairly
thin. Inflammatory cells are present ; the tubercle bacillus is seldom
found in urine in the absence of leucocytes.

In urine contaminated with *Mycobacterium smegmatis* (**146**) the
bacilli are thicker and individual cells are evenly stained. The bacilli are
scattered over the surface of epithelial cells. They do not remain attached
in clumps of parallel cells as do tubercle bacilli. Other non-acid-fast
bacteria are also attached to the epithelial cells. There are fewer leucocytes. (*Ziehl-Neelsen, x1,000.*)

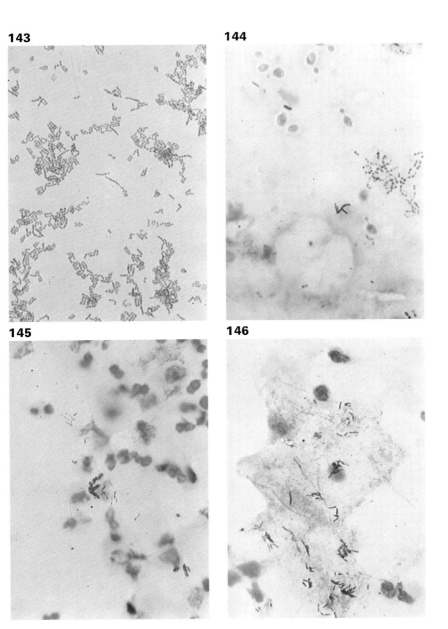

143

144

145

146

147 *Cord formation by the human tubercle bacillus* Mycobacterium tuberculosis under appropriate conditions grows in serpentine cords. The individual bacilli are arranged parallel to one another within a lipid matrix which can be removed with petroleum ether. This preparation was made from the water of condensation of a young growing culture on Löwenstein-Jensen medium. (*Ziehl-Neelsen, x 70.*)

148–152 *Smears from cultures of mycobacteria* The species cannot be determined by cellular morphology alone, but it provides useful confirmatory information which is readily available. Sometimes a stained smear is sufficient to exclude *M. tuberculosis*.

M. tuberculosis (**148**). The organism contains large clumps as well as smaller groups and single cells. In some places it can be seen that the bacilli within the clumps are arranged parallel to one another.

M. bovis (**149**). The field comprises well-dispersed single cells and small groups. The bacilli are shorter, slightly thicker and more evenly stained than those of *M. tuberculosis*. In preparing the smear the culture emulsified readily.

M. kansasii (**150**). The culture emulsified fairly well. The bacilli are long and unevenly stained. About 50% of strains have this appearance in smears; the others look more like *M. tuberculosis*.

M. xenopi (**151**). Characteristically the culture emulsifies well. The bacilli are long, slender and do not stain deeply. The arrangement of the cells often suggests a network (as here) or a skein.

M. avium (**152**). The organisms are evenly dispersed throughout the smear. The bacilli are characteristically short.

(*Ziehl-Neelsen, x1,000.*)

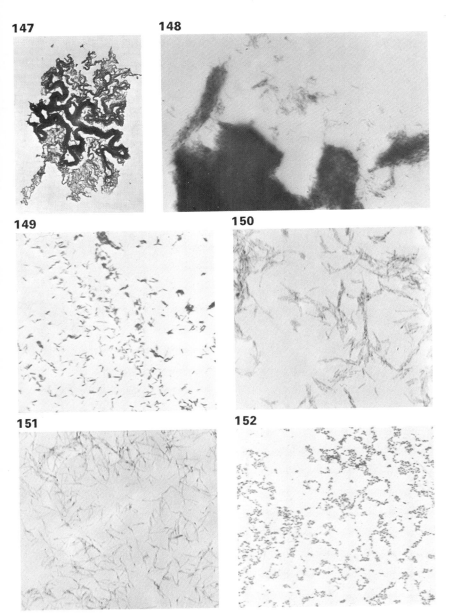

153 *Histological section of a skin nodule from a case of leprosy*
Bundles of red-stained leprosy bacilli are grouped inside mononuclear cells. In some cells the nucleus is absent, but the cell membrane is left surrounding the bacilli. In this way large globi may be formed (**154**). (*Triff, x1,000.*)

154 *Smear of the fluid squeezed from a leprous skin lesion* The giant globus at the bottom of the picture contains an enormous number of bacilli. In the smaller globus near the centre of the field the individual cells are more obvious. Even so they stain poorly and many of these bacilli were probably dead when the smear was prepared. (*Ziehl-Neelsen, x1,000.*)

155 *Partial acid-fastness of Nocardia asteroides* This smear from a culture shows typical branching filaments. It was stained by Ziehl-Neelsen except that 5% (rather than 25%) sulphuric acid was used for decolorisation. Some of the filaments are acid-fast by this treatment. (*Modified Ziehl-Neelsen, x1,000.*)

156 *Microcolonies of Actinomyces israelii* Young colonies are recognizable under the low power of the microscope by their rhizoid structure. For comparison, *A. bovis* produces a circular microcolony with an irregular edge ; the microcolony of *A. naeslundi* is somewhat intermediate between those of *bovis* and *israelii*. (*Anaerobic digest agar B with CO_2, 24 hours at 37°C; unstained, x170.*)

153

154

155

156

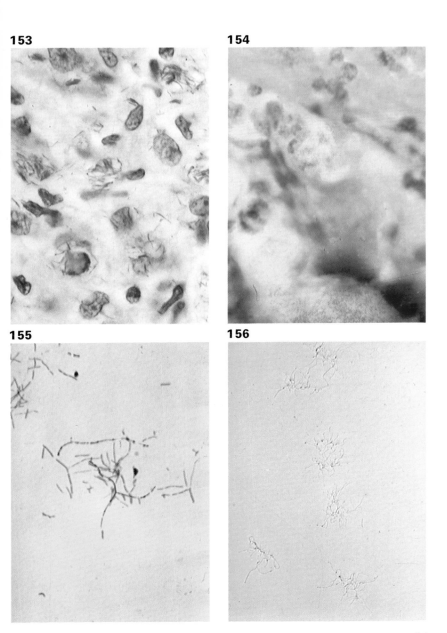

157 'Sulphur-granules' in pus from a case of actinomycosis
The pus was washed with distilled water and the solid sediment resuspended and photographed in a petri dish. The large granules are yellow
('sulphur-granules') ; smaller ones appear white. (*Unstained, x4.*)

158 Club-like structures on the surface of a sulphur granule
The granule was mounted in water and viewed through the microscope
with a reduced substage condenser aperture. Club-like structures radiate
from the edge of the granule. When one can focus up and down with
the microscope one appreciates that the entire surface of the granule
is covered by these so-called 'rays'. (*Unstained, x400.*)

159 A crushed 'sulphur-granule' stained by Gram The typical
gram-positive branching filaments, in this case of *Actinomyces bovis*,
may be demonstrated by pressing one of the smaller granules between
the slide and cover-slip, and staining the resulting smear. (*Gram, x1,000.*)

160 Smear from a culture of Actinomyces israelii The organisms
are more diphtheroid in culture, although some branching is evident —
perhaps rather more than usual in this field. (*Gram, x1,000.*)

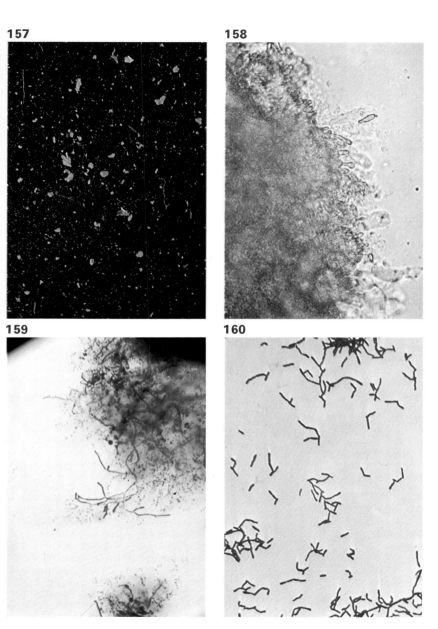

157

158

159

160

161 *Smear from a culture of Bacillus anthracis* Most of the cells
are arranged in chains, although there are a few singles and pairs. The
bacilli are large and gram-positive ; although this smear was prepared
from a young (18 hour) culture, a few cells have already lost the ability
to retain the gram stain. (*Gram, x1,000.*)

162 *Spores of Bacillus anthracis* This smear is from an older culture
than that shown in **161**. The spores are stained green and the vegetative
cells red. Some of the spores are still within the cells, but others are free.
Note that the cell bodies are not distended by the spores (cf. **166** and
167). (*Hot malachite green counterstained with safranin, x1,000.*)

163 *Bacillus anthracis in a blood smear from a guinea-pig* The
blue-stained bacilli are surrounded by pink capsules. This is the
M'Fadyean reaction. Experienced workers can diagnose anthrax even
when the capsule has disintegrated and is scattered between the bacilli.
The smear was fixed with Zenker's fixative, rather than heat, to kill the
anthrax spores. (*Polychrome methylene blue, x1,000.*)

164 *Smear from a culture of Bacillus cereus* Bacillus species
readily lose their gram-positive reaction. Gram-negative cells are
frequently found together with gram-positive cells. *Bacillus* species are
among the largest rods that one encounters in routine bacteriology.
(*Gram, x1,000.*)

161

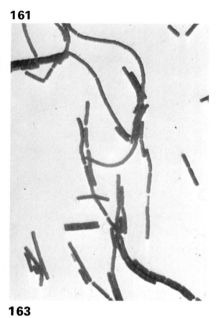

162

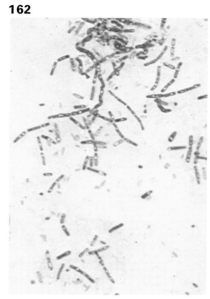

163

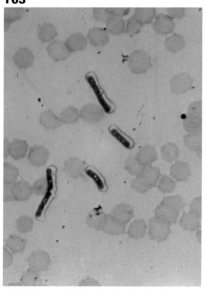

164

165 *Smear from a culture of Clostridium welchii* The morphology of this organism is sufficiently distinctive to enable experienced workers to distinguish typical strains from other clostridia. Such strains consist predominantly of single cells which are stout, relatively short, and have truncated ends and no spores. (*Gram, x1,000.*)

166 *Negatively stained smear of Clostridium tetani* The terminal spores when mature are spherical and of much greater diameter than the vegetative cells. This produces the 'drum-stick' morphology characteristic of this species. (*Nigrosin, x1,000.*)

167 *Smear of Clostridium sporogenes stained to show spores* Some of the cells show the bulge of the young spore ; others have a similar but larger swelling containing a green-stained spore. There are also a number of free spores. (*Hot malachite green counterstained with safranin, x700.*)

168 *Differential staining by fluorescent antibodies of two species of Clostridia* Fluorescent antibody methods allow rapid identification of bacteria and other micro-organisms in smears made either directly from infected tissues or from cultures. The specificity of these procedures is illustrated here. A smear containing a mixture of *C. septicum* and *C. chauvoei* was treated with an antibody against septicum labelled with fluorescein isothiocyanate and with one against chauvoei labelled with lissamine rhodamine B200. The identity of the individual cells of septicum (stained green) and chauvoei (red) is revealed in this preparation examined by fluorescence microscopy. (*x1,000.*)

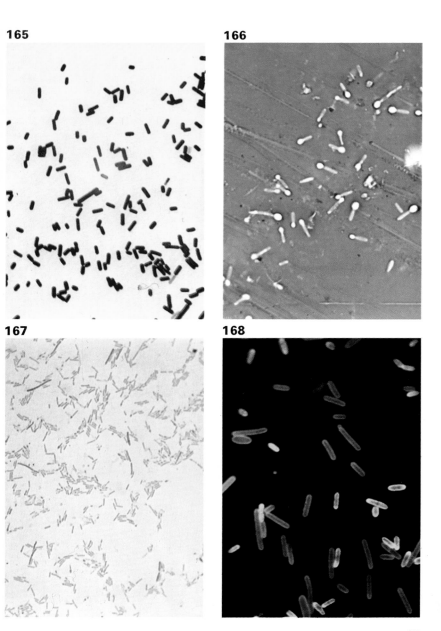

165

166

167

168

169 *A smear of Escherichia coli* Gram-negative rods of medium length and diameter ; single cells are evenly dispersed throughout the field. The cellular morphology of most members of the Enterobacteriaceae (see **69**) is indistinguishable from that of *E. coli* ; some of the exceptions are shown in **170** and **278**. (*Gram, x1,000.*)

170 *A smear of Klebsiella pneumoniae* Friedlander's bacilli are usually short and thick, and so they are frequently quite oval. They occur as single cells and as pairs evenly spread throughout the field. (Other capsulated coliforms frequently look like this too.) *K. pneumoniae* is more regular in the animal body ; here one finds short, stout diplobacilli surrounded by thick capsules. (*Gram, x1,000.*)

171 *A wet preparation to demonstrate capsules of Klebsiella pneumoniae* Growth from a plate culture was emulsified on a slide in India ink ; this was then compressed under a cover-slip. Each bacterial cell is surrounded by a halo, the capsule. The dense black material is India ink. There are also intermediate regions in which the ink is mixed with slime — less dense material which has become detached from the surface of the capsule. (*India ink, x1,000.*)

172 *Smear from a lesion of granuloma inguinale* There are groups of small bipolar rods in the cytoplasm of a large mononuclear cell. With other stains it can be shown that the organisms are capsulated and gram-negative. Cultural and antigenic studies suggest that it is a klebsiella which is highly adapted to a parasitic existence ; it requires special media for growth. (*Giemsa, x1,000.*)

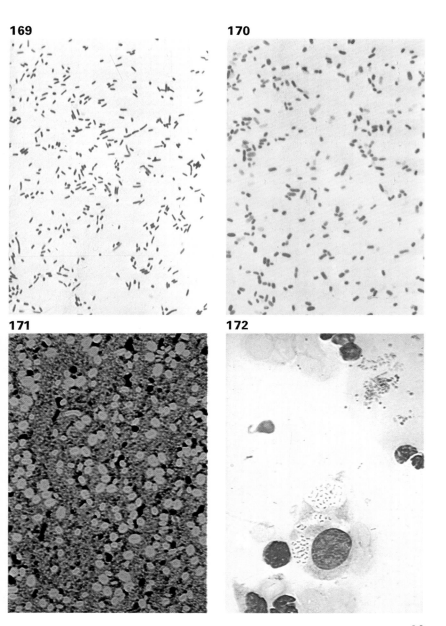

173 *Smear of Proteus vulgaris stained to demonstrate flagella*
Most of the bacterial cells are short. There are also a few very long forms
which have numerous peritrichate flagella; these are typical of the
swarming phase of *P. vulgaris*. All motile members of the Enterobacteriaceae have peritrichate flagella. (*Kirkpatrick's stain, x1,000.*)

174 *Smear of Pseudomonas aeruginosa stained to demonstrate
flagella* A single polar flagellum can be seen on each of the isolated
cells. (*Preston & Maitland's stain, x1,000.*)

175 *Smear of Yersinia pestis in a liver lesion* The numerous
plague bacilli show the bipolar staining characteristic of this species in
tissue preparations. Individual cells are shaped rather like safety pins.
Since some other gram-negative rods show bipolar staining, this appearance is not diagnostic of *Y. pestis*. (*Leishman, x1,000.*)

176 *Smear from the stomach of a bovine foetus showing
Brucella abortus* In-vivo grown *B. abortus* is usually associated with
tissue cells. It resists decolorisation with weak acid after being stained
by dilute carbol fuchsin. Some other less common pathogens stain
similarly. (*Cruickshank's modification of Koster's stain, x1,000.*)

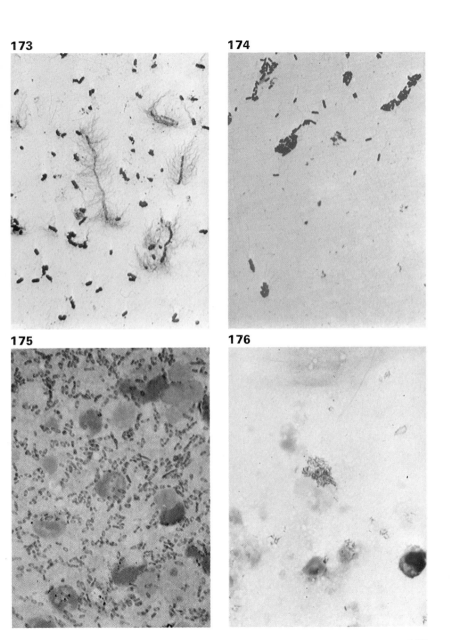

177 *Smear from a culture of Haemophilus influenzae* Most of the cells are tiny cocco-bacilli, but quite long forms are present also. Other strains may vary considerably from the typical appearance of *H. influenzae* shown here. (*Gram, x1,000.*)

178 *Smear of Haemophilus parahaemolyticus* Compare these large tangled filaments with the tiny bacilli of *H. influenzae* shown in **177**. *H. parahaemolyticus* dies out within a few days on a blood agar plate. Degenerating forms are obvious at 24 hours; they are pale, granular and swollen into globular or moniliform shapes (cf. **180**). (*Gram, x1,000.*)

179 *Smear from a culture of a moraxella* These are gram-negative rods which typically occur in pairs. Short thick well-stained pairs are seen together with poorly-stained longer forms. Some strains are even more pleomorphic than that shown here; 'ghost-cells' and partially gram-positive cells may be found as well. (*Gram, x1,000.*)

180 *Smear from a 48 hour blood agar culture of Streptobacillus moniliformis* There are many interlacing filaments or chains of rods. In some places the filaments are swollen into L-forms, (large spindle or pear-shaped forms, some of which show internal granules). A smear from a younger culture would show fewer L-forms. (*Gram, x1,000.*)

177

178

179

180

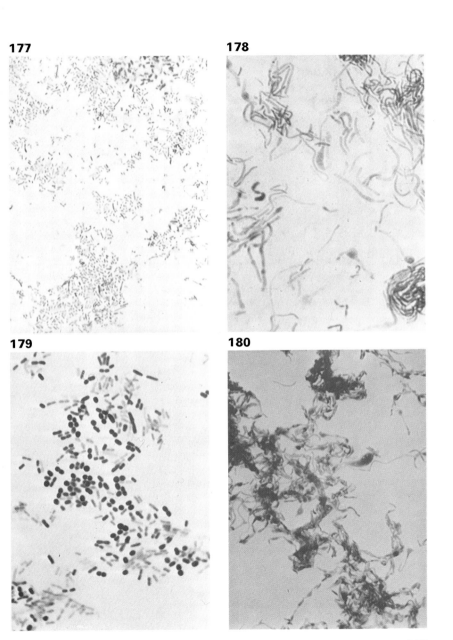

181 *Granule from a culture of Streptobacillus moniliformis*
The bulbous expansions around the periphery of the granule are helpful in the recognition of this species. (*Serum broth, 48 hours at 37°C; unstained, x280.*)

182 *Mycoplasma suipneumoniae* Seen here in an impression preparation from a pig's lung. Tiny coccoid, bipolar and ring forms can be seen scattered among the tissue cells. (*Giemsa, x1,000.*)

183 *Smear from a culture of Campylobacter fetus* This organism is a vibrio in young cultures, but later produces the spirillar forms shown. It is a well-known cause of abortion in sheep and cattle. A few instances of human infection with this and similar organisms are also on record. (*Gram, x1,000.*)

184 *A blood smear from a mouse showing Spirillum minus* The mouse had been inoculated with blood from a human case of rat bite fever. The helical organism in the middle of the field is the spirillum ; it is usual to find not more than one organism per field (cf. **186**). *S. minus* is a commensal of the mouth of the rat. It is transmitted to man by the bite of a rat ; the resulting disease is one of the two forms of rat bite fever. (*Leishman, x1,000.*)

181

182

183

184

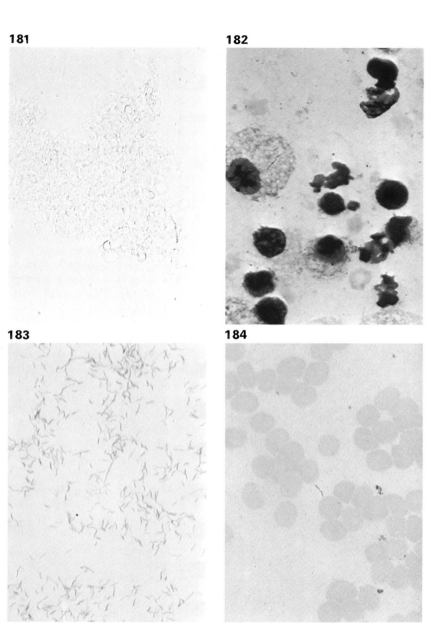

185 *Smear from a lesion of Vincent's angina* *Borrelia vincenti* and *Fusobacterium fusiforme* were numerous in the preparation. The borrelia is a large spirochaete with up to ten coils. The fusobacterium is a large, barred, non-motile rod with pointed ends. Since both organisms are found in the healthy mouth it is not clear whether they cause the lesions of Vincent's angina or are secondary invaders. (*Dilute carbol fuchsin, x1,000.*)

186 *Borrelia recurrentis in a smear of mouse blood* The mouse had been inoculated with the blood of a patient in the pyrexial phase of relapsing fever. The borrelia has 4–6 coils, each of which may be up to 3μm long. The clumping of the organisms indicates that agglutinating antibody has appeared in the mouse blood. Similar forms, but in smaller numbers, may be found in the blood of a patient during the pyrexial phase of the disease. (*Giemsa, x1,000.*)

187 *Smear from a culture of Leptospira icterohaemorrhagiae* These extremely thin organisms are rendered visible by the precipitation of silver stain on their surfaces. They have very fine spirals and distinctive terminal hooks. (*Fontana, x1,000.*)

185

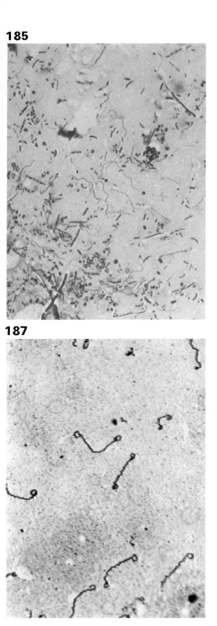

186

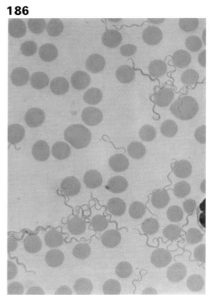

187

Fungi: Macroscopic Appearances

In this and the following chapter we deal first with the dermatophytes, which attack keratinised skin, hair and nails, and then with the fungi which infect deeper tissues. The dermatomycoses are common and are quite contagious; many of them are transmissible between animals and man. They are commonly called ringworm or tinea.

The mycoses of deeper tissues are relatively rare and are not readily transmitted from one person to another. They are caused by fungi which invade through the skin, or the alimentary tract, or in most species through the lungs. Some of them may become widely disseminated throughout the body. Some of these fungi, such as *Cryptococcus neoformans*, are found all over the world. Others are restricted to certain geographical regions where the soil is suitable for their saprophytic existence; with the increasing frequency of international travel, cases of the latter group are becoming more widespread. *Aspergillus fumigatus* and *Candida albicans* are widely distributed, and are common opportunist pathogens in patients whose normal flora is disturbed by antibacterial agents, or in those whose resistance is lowered by immuno-suppressive drugs.

Whereas bacterial colonies are studied on agar plates, with fungi cultures in cotton-wool-plugged test tubes or screw-capped bottles are usually preferred. These containers are safer, suffer less cross-contamination, and are more suitable for prolonged incubation under the aerobic conditions required by fungi. Therefore most of the photographs in this chapter are of tube cultures. Since they are usually examined with the naked eye they are reproduced unmagnified; most show front and rear views.

The macroscopic morphology, and particularly the pigmentation, of the growth on Sabouraud's glucose agar is particularly useful in the identification of species. Most of the specimens shown had been incubated at 26° and some at 37°C.

188 *Tube cultures of Microsporum canis* The growth is white and cottony. The reverse, right, is bright yellow, with a deeper orange under the areas of denser growth. (*Sabouraud glucose agar, 4 weeks at 26°C; reflected + transmitted light, x1.*)

189 *Tube cultures of Microsporum audouinii* The colony is flat and has a fine felt texture, and a central boss. (*Sabouraud glucose agar, 12 days at 26°C; reflected light, x1.*)

190 & 191 *Tests for growth of Microsporum spp. on rice grains*
M. canis (**190**) produces a granular growth over the surface of the rice. It is coloured tan, and the substrate is pinkish buff. *M. audouinii* (**191**) fails to grow. Although the rice grains at the site of inoculation are discoloured brown, no mycelium is visible. (*Sterile polished rice grains, 10 days at 26°C; reflected light, x½.*)

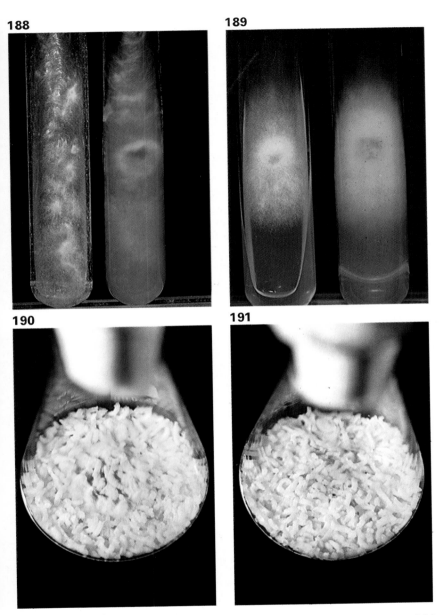

192 *Tube cultures of Microsporum gypseum* The colony has a white sporeless border enclosing a pale, fawn coloured area with a texture somewhat like chamois leather. The reverse is pale yellow. (*Sabouraud glucose agar, 7 days at 26°C; reflected light, x1.*)

193 *Tube cultures of Trichophyton mentagrophytes, granular type* *Trichophyton mentagrophytes* is commonly carried by animals, and strains derived from an animal source usually have a granular surface. This results from an abundance of microconidia. These are readily spilled when making culture transfers, so that the growth originates from many foci. This is obvious in the reverse tube, right; each focus has a light yellow reverse with a denser tan centre. (*Sabouraud glucose agar, 7 days at 26°C; reflected light, x1.*)

194 *Tube cultures of Trichophyton mentagrophytes, downy variety* Cultures grown from tinea pedis have few microconidia and are filamentous. The reverse is red-brown with a lighter periphery. This type is also called *Trichophyton interdigitale.* (*Sabouraud glucose agar, 4 weeks at 26°C; reflected light, x1.*)

195 *Tube cultures of Trichophyton rubrum* This species usually produces a white cottony colony. A characteristic wine-red reverse develops as the colony ages. (*Sabouraud glucose agar, 4 weeks at 26°C; reflected light, x1.*)

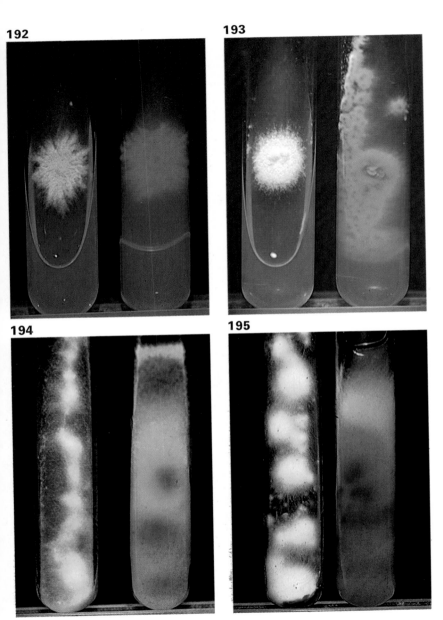

196 *Tube cultures of Trichophyton verrucosum* This species grows very slowly, especially when first isolated. The tube on the left shows the amount of growth to be expected from a primary culture on malt extract agar (*2 weeks at 26°C*). Growth in subculture is better, even on simpler media. The tube on the right shows a subculture on Sabouraud glucose agar (*4 weeks at 26°C*). The colonies are at first waxy; later they become white, heaped and irregularly and deeply folded (*reflected light, x1*).

197 *Tube cultures of Trichophyton sulphureum* This is a slowly growing species which, in the early stage shown here, is flat and has a powdery surface. It produces a distinctive sulphur-yellow pigment which can be seen in both surface and reverse views. The colony later became heaped and folded. (*Sabouraud glucose agar, 2 weeks at 26°C; reflected light, x1.*)

198 *Tube cultures of Trichophyton violaceum* The purple pigment from which this species derives its name is most concentrated in the irregularly folded centre of the colony. This is surrounded by a radially folded zone. (*Malt extract agar, 4 weeks at 26°C; reflected light, x1.*)

199 *Tube cultures of Trichophyton violaceum on Sabouraud glucose agar* This is the same strain as that shown in **198**. On this medium the distinctive pigment is lacking, and variant cottony sectors arise which overgrow the colony. This is a good example of how colony appearance depends on the culture medium. (*Sabouraud glucose agar, 2 weeks at 26°C; reflected light, x1.*)

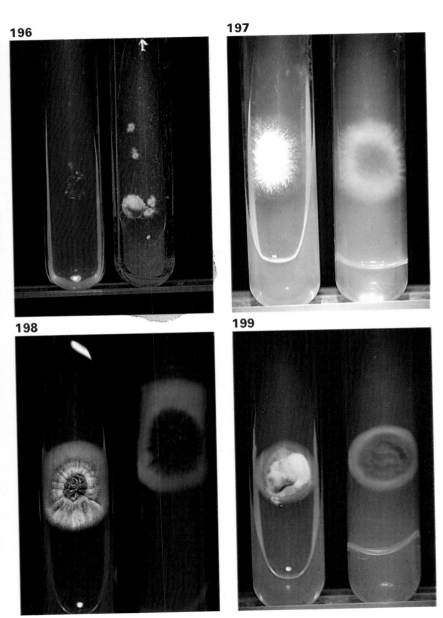

200 *Tube cultures of Trichophyton schoenleini* The white leathery colony becomes deeply folded, and cracks develop along the tops of the folds. Eventually the cracks deepen and the underlying medium splits with them. (*Sabouraud glucose agar, 2 weeks at 26°C; reflected light, x1.*)

201 *Tube cultures of Epidermophyton floccosum* This young colony is thin and flat, and has a distinctive greenish-yellow colour which is apparent on both the obverse and reverse. Later the colony becomes folded and the pigment becomes more apricot-orange on the reverse side. (*Sabouraud glucose agar, 8 days at 26°C; reflected light, x1.*)

202–204 *Some other examples of dermatophyte colonies* For each dermatophyte species one can expect a range of cultural types, and it may be misleading to give only one example of each range. Three other examples are included here for comparison with those illustrated previously. (*4 weeks at 26°C; reflected light, x1.*)

Figure **202** shows the depth of pigment obtainable in a well grown culture of *Trichophyton rubrum*. (*Sabouraud glucose agar.*)

A *Trichophyton violaceum* colony (**203**) which has a downy surface and no radial grooves. Note how the pigment is visible from the surface of the colony and is most concentrated under its centre. (*Malt extract agar.*)

This *Trichophyton schoenleini* colony (**204**) is coloured tan. It shows a rich mycelial growth into the medium. This is why the mycelium becomes so involved in the cracking of the folds of the medium. (*Sabouraud glucose agar.*)

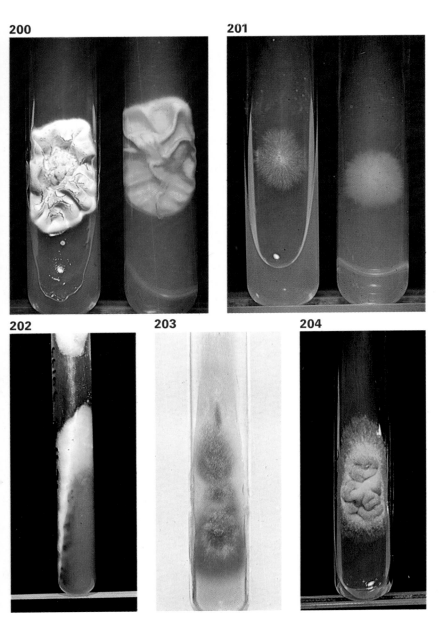

205 *Yeasts growing in pure culture from the faeces of a patient who had received a broad-spectrum antibiotic* By killing off the bacterial flora of the gut the antibiotic has enabled the yeast to proliferate ; this itself then caused alimentary disturbance. The colonies are whitish, domed and have a fine matt surface. Species identification of the yeast was not attempted. (*Blood agar A, 24 hours at 37°C; reflected light, x6.*)

206 *Colonies of Candida albicans growing from a vaginal swab* A routine medium for plating vaginal swabs is heated blood agar incubated in an atmosphere of carbon dioxide. Many *C. albicans* strains develop a very characteristic morphology under these conditions. There are numerous filamentous projections ('whiskers') of pseudomycelium growing from the yeast-like colony. (*Heated blood agar A, 24 hours at 37°C; reflected light, x6.*)

207 *Two tube cultures of Cryptococcus neoformans grown at different temperatures* Growth is equally profuse at 26°C, left, and at 37°C, right. The luxuriant, creamy, mucoid colony has run down the slopes. (*Sabouraud glucose agar; reflected light, x1.*)

208 *Colonies of Cryptococcus neoformans on blood agar* Some of the mucoid colonies have white opaque sectors, alternating with clearer areas. (*Blood agar B, 5 days at 37°C; reflected light, x6.*)

205

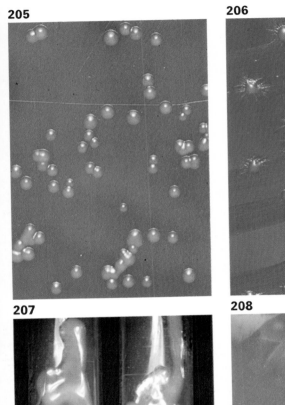

206

207

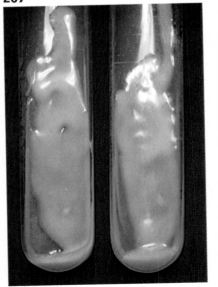

208

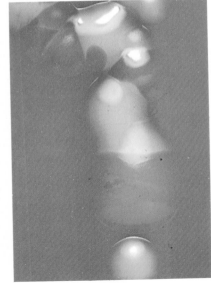

209 *Yeast phase culture of Blastomyces dermatitidis* In the identification of dimorphic fungi, it may be necessary to demonstrate both phases in culture. These yeast-phase colonies are waxy, raised and are already showing some surface irregularities. (*Heated cysteine glucose blood agar, 2 weeks at 37°C; reflected light, x1.*)

210 *A yeast phase culture of Sporotrichum schenckii* Any of the dimorphic fungi may produce yeast-phase cultures which look like this. However, microscopic preparations from such cultures are most helpful in identification of the culture. (*Heated cysteine glucose blood agar, 7 days at 37°C; reflected light, x1.*)

211 *Young mycelial phase culture of Sporotrichum schenckii* The shiny, white, glabrous colony soon develops radial folds and a fimbriate edge. (*Sabouraud glucose agar, 8 days at 26°C; reflected light, x1.*)

212 & 213 *Giant colonies of two blastomycosis organisms* On prolonged incubation these species grow into large waxy colonies of characteristic appearance. The isolated wrinkled colony of *Blastomyces dermatitidis* (**212**) is so tall that its top is out of focus. The single colony of *Paracoccidioides brasiliensis* (**213**) near the edge of the culture medium is rather reminiscent of a worm cast. (*Blood agar B, 5 weeks at 37°C; reflected light, x4.*)

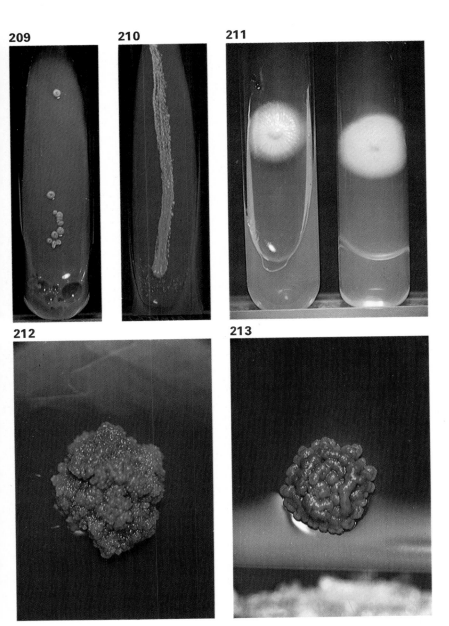

214 *Mycelial phase culture of Blastomyces dermatitidis* The colony is smooth at first, but then becomes white and cottony (shown here). It may later become tan coloured. Growth of *Paracoccidioides brasiliensis* is similar to this. (*Sabouraud glucose agar, 4 weeks at 26°C; reflected light, x1.*)

215 *Mycelial phase cultures of Histoplasma capsulatum* This slowly-growing culture was white at first, and later became coloured light brown as shown. (*Sabouraud glucose agar, 4 weeks at 26°C; reflected light, x1.*)

216 *Culture of Coccidioides immitis* A white, woolly growth with some dark-tinged areas. Other strains may look like this, but have a clumped central zone. Later the mycelium will fragment to show a dry, crumbly growth from which clouds of highly infectious spores could be liberated. Because of the hazard this entails, *C. immitis* should be handled only in bottles and in specially equipped laboratories. (*Sabouraud glucose agar, 4 weeks at 26°C; reflected light, x1.*)

217 *An older culture of the strain of Sporotrichum schenckii shown in* **211** The tough, membranous colony has developed brownish-black areas. The radial furrows have deepened, and the centre has become irregularly folded. (*Sabouraud glucose agar, 4 weeks at 26°C; reflected light, x1.*)

214

215

216

217

218 *Tube culture of Phialophora sp.* All species are dark-grey to black on both surfaces. The colony is deeply embedded in the medium. The colony shown is *Ph. verrucosa.* However the species cannot be differentiated on colony morphology ; for this, microscopic examination of spore structures (**254** and **255**) is required. (*Sabouraud glucose agar, 14 days at 26°C; reflected light, x1.*)

219 *Tube cultures of Aspergillus niger* Beginning at the centre, black spore-heads visible to the naked eye develop on the flat, rapidly growing colony. The reverse of the colony is white ; this helps to differentiate *Aspergillus niger* from other black fungal colonies. Radial furrows can be seen from either surface. (*Sabouraud glucose agar, 4 days at 26°C; reflected light, x1.*)

220 *Tube cultures of Aspergillus fumigatus* The structure of the colony is similar to that of *Aspergillus niger,* except that the spore heads are smaller, and are coloured smoky blue-green, later becoming darker. (*Sabouraud glucose agar, 4 days at 26°C; reflected light, x1.*)

221 *Tube cultures of Penicillium sp.* A flat colony with radial furrows, and with a grey-green powdery surface. Some strains exude yellow droplets on to the surface of the colony. (*Sabouraud glucose agar, 4 days at 26°C; reflected light, x1.*)

218

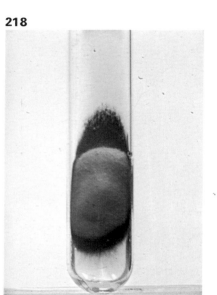

219

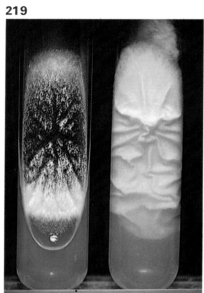

220

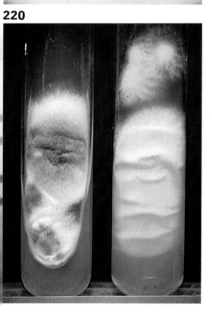

221

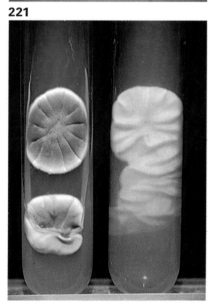

Fungi: Microscopic Appearances

Morphological features are most helpful in identifying fungi. The use of other procedures such as biochemical and immunological tests is increasing; that these methods have been applied more to bacteria is indicative of the relatively limited value of morphological examination of bacteria as compared with fungi. The macroscopic examination of cultures as illustrated in the previous chapter is usually supplemented by microscopy.

Material for microscopic examination for fungi may be prepared directly from a lesion or from a growing culture. Hairs or scrapings from the edge of a skin lesion may be cleared and examined immediately, as may stained or unstained smears of pus. Many systemic fungal diseases are diagnosed by examination of histological sections, usually because material for culture was not taken before the lesion was treated with fixative. Photographs of histological sections are not included in this chapter except for the non-cultivable *Rhinosporidium seeberi*.

Material taken from a growing culture may be mounted in lactophenol-cotton blue which clears and stains the fungal elements. However this readily disturbs the arrangement of the diagnostic structures. Better preparations for microscopy are made with slide cultures from which the agar block can be removed to leave the fungal elements undisturbed on the slide or cover slip.

For most of the pathogenic fungi only the imperfect (asexual) forms are seen in the routine laboratory. Moreover the growth medium and temperature markedly influence the organism's appearance. A dermatophyte growing in hair produces hyphae which break down into arthrospores of characteristic location and arrangement. When it grows on an appropriate agar medium the organism may produce different types of spores and other structures which are useful in identification.

Many species may be recognized by the frequency, grouping and shape of their large, multiseptate macroconidia; the three genera of dermatophytes, *Microsporum, Trichophyton* and *Epidermophyton*, are distinguished primarily by the morphology of their macroconidia. The microconidia are unicellular; again, their shape, frequency, location and arrangement are distinctive for some species. Other structures such as chlamydospores, and spiral hyphae are typical of certain species.

The so-called dimorphic fungi may produce mold-like growth at room temperature on Sabouraud's glucose agar, and yeast-like growth at 37°C on blood agar. Still other structures may be produced *in vivo*.

Cultures of some of these species, especially *Coccidioides immitis*, release highly infectious airborne spores which constitute a serious hazard to laboratory workers.

222–225 *Microscopic examination of hairs attacked by dermatophytes* When dermatophytes grow on hair, their mycelium breaks down into arthrospores. Hair may be examined directly for these refractile spores by mounting it in a clearing agent such as potassium hydroxide, lactophenol or xylol. Some indication of the identity of the causative fungus may be obtained by noting the size of the spores (small or large), whether they are arranged in chains or scattered irregularly to form a mosaic, and whether they are inside the hair shaft (endothrix) or outside (ectothrix). (*Hairs cleared in lactophenol, x200.*)

222, a mosaic ectothrix infection. Small spores are irregularly disposed outside the hair. This arrangement is characteristic of *Microsporum* species.

223, large-spore ectothrix. Although there are spores outside the hair, they are inside as well and the hair is greatly expanded by fungal growth (only about half the diameter of the hair is shown). This is sometimes referred to as an endo-ectothrix infection; it is characteristic of some *Trichophyton* species.

224, linear small-spore ectothrix infection. The chains of arthrospores have been formed from mycelium growing on the surface of the hair. This appearance may be found in *Trichophyton* or *Microsporum* infections. (*Hair mounted in lactophenol-cotton blue.*)

225, endothrix infection. That the spores are within the hair shaft is most readily appreciated when one can focus up and down with the microscope. However it is clear that there are virtually no spores outside this hair – the microscope is focused on the widest diameter. Chains of spores are most obvious at the top of the picture.

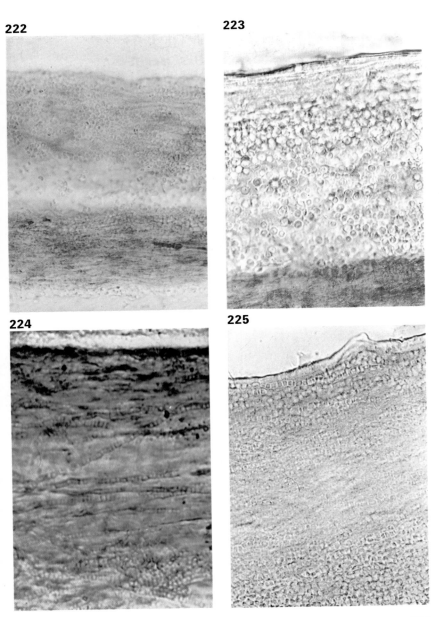

222

223

224

225

129

226 *A normal hair* (*above*) *compared with one infected with Trichophyton schoenleini* This fungus, the causative agent of favus, grows within the hair but seldom shows arthrospores. Where the hyphae have grown within the shaft, they leave distinctive refractile spaces. (*Lactophenol, x200.*)

227 *Penetration of hair by Trichophyton mentagrophytes in vitro* The organism was cultured on Sabouraud's agar on which were laid a few pieces of blond hair. The fungus has invaded the hair by tunnelling perpendicular to its surface. *T. rubrum* does not produce such perforations *in vitro*. (*Sabouraud glucose agar, 4 weeks at 26°C; unstained, x70.*)

228 *Slide culture of Microsporum canis* The macroconidia are large and fusiform with pointed ends and thick walls. They contain 6–12 cells and have a verrucose surface. (*Slide culture on rice grains, 7 days at 26°C; lactophenol-cotton blue, x170.*)

229 *Slide culture of Microsporum gypseum* There are many macroconidia which contain 4–6 cells. They are spindle-shaped, but are more rounded than those of *M. canis,* and their walls are thinner. There are also a number of clavate (club-shaped) microconidia. (*Sabouraud glucose agar, 7 days at 26°C; lactophenol-cotton blue, x170.*)

226

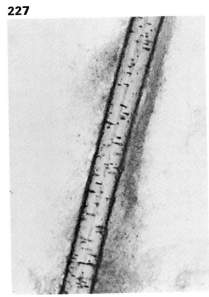

227

228

229

230 *Slide culture of Microsporum audouini* Although some strains have microconidia resembling those of other *Microsporum* species, the strain shown has none. Macroconidia are also rare in this species, and when present they are usually ill-formed. In the centre of the field is shown the most distinctive microscopic feature of this species, a terminal chlamydospore with a pointed tip. (*Sabouraud glucose agar, 7 days at 26°C; lactophenol-cotton blue, x280.*)

231 *Slide culture of Trichophyton mentagrophytes* The granular appearance of the colony of this strain (**193**) results from its numerous globose microconidia, which are born in large clusters ('en grappe'). However this field was selected for its examples of spiral hyphae, some of which are tightly coiled. Only *T. mentagrophytes* may have so many spiral hyphae, although some strains of this and other dermatophytes may have a few. (*Sabouraud glucose agar, 7 days at 26°C; lactophenol-cotton blue, x170.*)

232 *Needle mount from a culture of Trichophyton mentagrophytes* Macroconidia are rare in *Trichophyton* species; this field has been selected to show their morphology when present. They are elongated, smooth and thin-walled and have 2–6 cells. In some strains they are not much thicker than the hyphae, from which they are then not readily distinguished. (*Sabouraud glucose agar, 10 days at 26°C; lactophenol-cotton blue, x170.*)

233 *Slide culture of Trichophyton rubrum* Many clavate or pyriform microconidia are attached along the sides of the hyphae ('en thyrse'); this gives an appearance of tangled barbed wire. There is a single three-celled macroconidium (centre) which has smooth, parallel sides and a rounded tip. Spiral hyphae are rare in this species. (*Sabouraud glucose agar, 2 weeks at 26°C; lactophenol-cotton blue, x170.*)

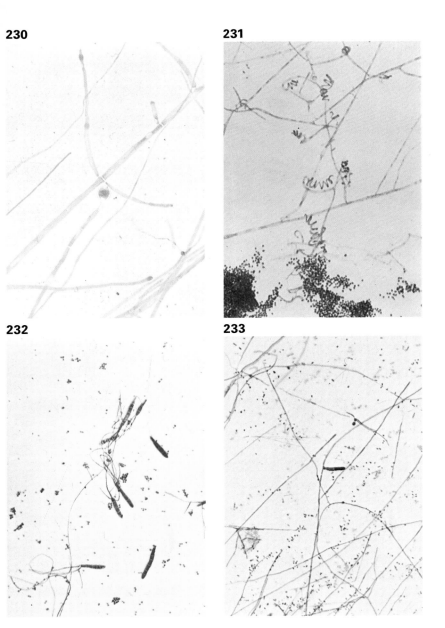

230

231

232

233

234 *Slide culture of Trichophyton verrucosum* The distinctive feature of this species on Sabouraud glucose agar is its irregular mycelium which breaks down into chains of chlamydospores. Macro and microconidia are very rare on this medium; they require enriched medium for their production. (*Sabouraud glucose agar, 7 days at 37°C; lactophenol-cotton blue, x280.*)

235 *Slide culture of Trichophyton schoenleini* This species produces no macroconidia, and microconidia are rare. Its mycelium is very irregular and in places the hyphae expand into bulbous ends. When there are multiple bulbs on a hypha they may resemble a moose antler. The irregular expansions of the hypha are also called 'favic chandeliers'; these are shown here. (*Sabouraud glucose agar, 14 days at 26°C; lactophenol-cotton blue, x280.*)

236 *Slide culture of Epidermophyton floccosum* This species is recognized by its short club-shaped macroconidia, which occur in groups. They have 2–4 cells and smooth walls. There are no microconidia. (*Sabouraud glucose agar, 14 days at 26°C; lactophenol-cotton blue, x170.*)

237 *An older slide culture of Epidermophyton floccosum* This field contains many chlamydospores. Some are terminal; others are within the length of the hyphae (intercalary). (*Sabouraud glucose agar, 18 days at 26°C; lactophenol-cotton blue, x170.*)

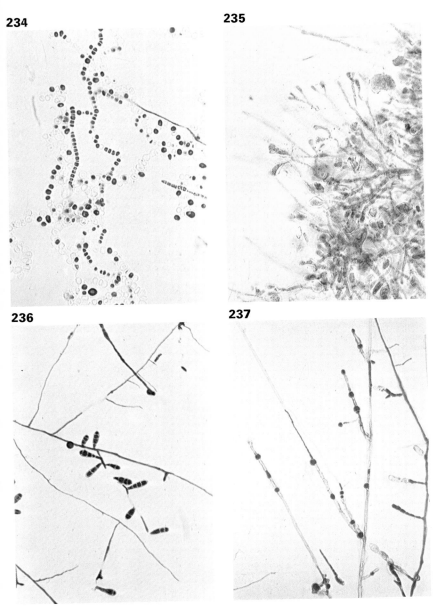

234

235

236

237

238 *Smear of a yeast in faeces* The gram reaction is rather variable. There are a few oval, purple buds, but the organism is chiefly in the mycelial phase. This is associated with some degree of invasion of host tissue. The pathogenicity of the normal yeast flora of the gut is commonly associated with treatment of the patient with a 'broad-spectrum' anti-bacterial drug. This patient had such a history. (*Gram stain, x700.*)

239 *Smear of Candida albicans prepared from a culture* The preparation consists entirely of strongly gram-positive oval yeasts which are budding in places. Yeast colonies may be confused with bacterial colonies, but the difference in size of the two groups of organisms is most obvious under the microscope. This photograph is reproduced at the same magnification as the earlier photomicrographs of bacteria. (*Gram stain, x1,000.*)

240 *Growth of Candida albicans* Seen here in the depths of corn-meal agar, examined through the base of the petri dish. Clusters of blastospores are borne at some joints of the pseudomycelium, and there are terminal, large, round chlamydospores. (*Cornmeal agar, 3 days at 26°C; unstained, x170.*)

241 *Formation of germ tubes by Candida albicans* If colonies of *C. albicans* are emulsified in human, horse or rabbit serum, and incubated at 37°C for 3 hours, germination of the yeast cells can be seen under the microscope. The germ tubes are thinner, but may be much longer than the parent cells. This provides a rapid identification of *C. albicans*. (*Horse serum, 3 hours at 37°C; phase contrast, x280.*)

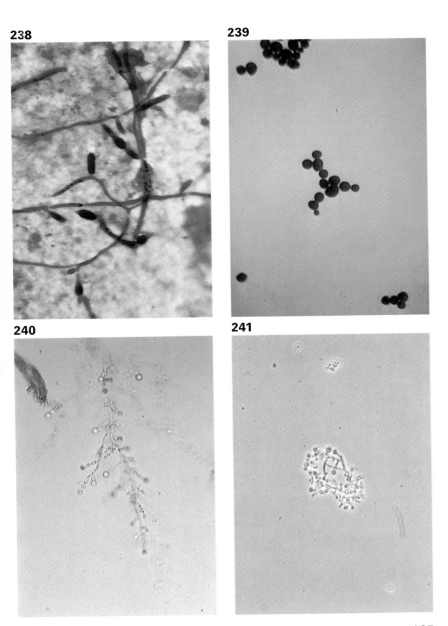

242 *Cryptococcus neoformans in mouse brain* The animal died six days after intracerebral infection. A small piece of brain was teased out on a slide in India ink, and then compressed under a cover slip. This procedure demonstrates the yeast cells, with their characteristic capsules outlined by the ink. A bud can be seen on the cell near the centre of the field. (*India ink, x170.*)

243 *Mycelial phase of Blastomyces dermatitidis* The conidia are round or pyriform ; some are sessile, but most are borne terminally on short simple conidiophores. (*Sabouraud glucose agar, 4 weeks at 26°C; lactophenol-cotton blue, x280.*)

244 *The yeast phase of Blastomyces dermatitidis* Large, thick-walled cells which produce broad-based buds. Similar forms may be found in infected animals. (*Blood agar B, 4 weeks at 37°C; unstained, x280.*)

245 *Yeast form of Paracoccidioides brasiliensis* It is seen in a pus smear from the testicle of an experimentally infected guinea-pig. Some cells may produce single buds, others moniliform chains, and still others the characteristic multiple buds. Their constricted point of attachment further differentiates them from the broad-based buds of *Blastomyces dermatitidis* (**244**). Similar forms may be found in yeast phase cultures. (*Three weeks after infection, Gridley stain, x280.*)

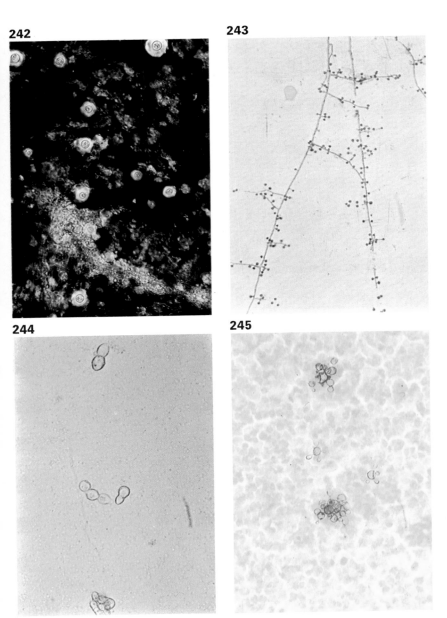

242

243

244

245

246 *Mycelial phase of Histoplasma capsulatum* The macroconidia are large, spherical and thick-walled ; they bear large tubercles.
Round to pyriform microconidia are also present. (*Sabouraud glucose
agar, 4 weeks at 26°C; faded lactophenol-cotton blue, x280.*)

247 *Mycelial form of Sporotrichum schenckii* Rosettes of oval
or pyriform conidia at the ends, and sometimes down the sides, of delicate conidiophores. Some of the vegetative mycelium is arranged in
parallel bundles. (*Sabouraud glucose agar, 4 days at 26°C; lactophenol-
cotton blue, x280.*)

248 *Yeast phase of Sporotrichum schenckii* This is a wet preparation from a culture like that shown in **210**. It consists predominantly
of yeast-like cells which are oval or elongated into the cigar shapes
characteristic of this species. Similar forms are found *in vivo*. (*Blood
agar, 7 days at 37°C; phase contrast, x280.*)

249 *Asteroid body of Sporotrichum schenckii* *S. schenckii* occurs
in the natural disease as a budding yeast, and more rarely as an asteroid
body as shown here, in which the yeast is surrounded by radiating
eosinophilic projections. (*Haematoxylin and Eosin, x700.*)

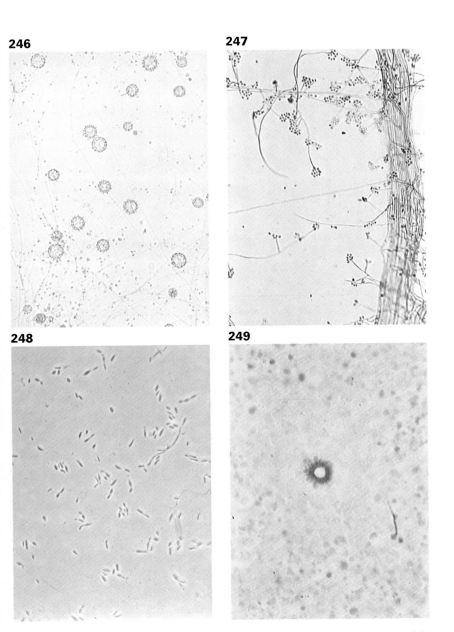

250 *Arthrospores of Coccidioides immitis* The mycelium forms chains of barrel-shaped arthrospores which alternate with unstained empty cells. When these arthrospores are released by fragmentation of mycelium, they retain cell fragments at either end, which assists their airborne dispersal. It is this that makes *C. immitis* so hazardous in the laboratory. (*Sabouraud glucose agar, 4 weeks at 26°C; lactophenol-cotton blue, x280.*)

251 *Needle mount from a culture of Aspergillus niger* The large spore-heads have remained dense, black and relatively unbroken in preparing the specimen. They are borne on long non-septate conidiophores. This species has been isolated from ear disease. (*Sabouraud glucose agar, 4 days at 26°C; lactophenol-cotton blue, x70.*)

252 *Slide culture of Aspergillus fumigatus* The conidiophore terminates in a flask-shaped vesicle which bears a row of parallel sterigmata from which chains of spherical conidia are produced. (*Sabouraud glucose agar, 4 days at 26°C; lactophenol-cotton blue, x170.*)

253 *Slide culture of Penicillium sp.* In this case short branches of the conidiophore bear the sterigmata which produce the chains of conidia. The structure has been likened to a brush (penicillus) from which the genus is named. (*Sabouraud glucose agar, 4 days at 26°C; lactophenol-cotton blue, x170.*)

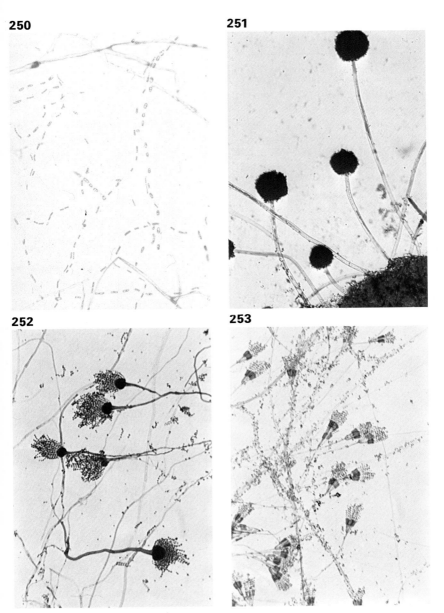

250

251

252

253

254 *Cladosporium-type conidiophores in a slide culture of Phialophora pedrosoi* The conidia are borne singly or in chains on the ends of branching conidiophores of varying complexity. (*Sabouraud glucose agar, 2 weeks at 26°C; lactophenol-cotton blue, x280.*)

255 *Phialophora-type conidiophores in a slide culture of Phialophora verrucosa* Each conidiophore is shaped like a flask with a flared lip. The spores are budded from the base of the conidiophore, and accumulate around its mouth. (*Sabouraud glucose agar, 2 weeks at 26°C; lactophenol-cotton blue, x280.*)

256 *Slide culture of Mucor sp.* This is a typical phycomycete. The hyphae are non-septate and of comparatively large diameter. The asexual spores are contained within a sac (sporangium) borne on a long sporangiophore. Sexual spores also may be found in cultures; they are produced by the fusion of two hyphae. Some phycomycetes are common culture contaminants, and some occasionally cause severe lesions in man and animals. (*Sabouraud glucose agar, 7 days at 26°C; lactophenol-cotton blue, x70.*)

257 *Section through a polyp caused by Rhinosporidium seeberi* There are a number of spherical sporangia, including (centre right) a large one which has a thick, refractile wall, and which contains many endospores. This organism has never been successfully cultivated, but that it is a fungus is suggested by the presence of cellulose and of chitin in its wall, and by some similarities between its sporangia and those of *Coccidioides immitis*. It is found in polyps, usually in the nose, of man and domestic animals. (*Haemotoxylin and Eosin, x170.*)

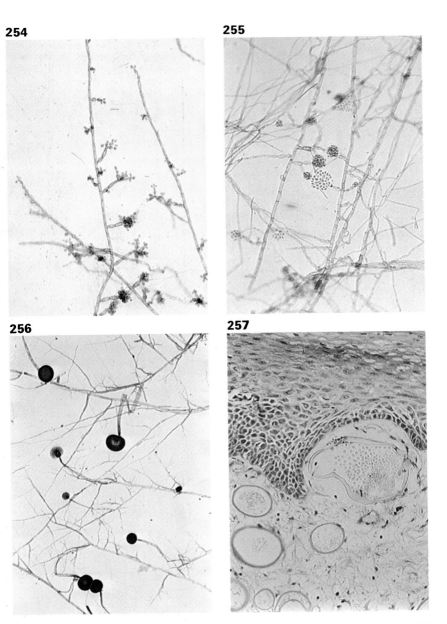

254

255

256

257

Antibiotics and Chemotherapeutic Agents

The photographs in this section illustrate some of the actions of anti-biotics and other chemotherapeutic agents most frequently seen in the laboratory. Methods are shown of assaying antibiotics in small volumes of liquids. Some illustrations give clues to the mode of action of these agents ; others to the mode of development of resistant variants.

In general, within any one species one should expect different strains to have different sensitivities to chemotherapeutic drugs. For example, different strains of *Staphylococcus aureus* differ greatly in their sensitivities to the commonly used drugs.

On the other hand, in some species the susceptibilities of all strains are sufficiently uniform to be helpful in identification. Antibiotic sensitivity tests are usually done on primary cultures from clinical material. This enables the microbiologist to give early information about the treatment of the patient. It also provides early information about the organism's identification. Thus the laboratory worker will regard a certain sensitivity pattern as a means of recognition of certain species. Their sensitivity pattern (or 'antibiogram' as it has been called) should prove useful in the identification of staphylococci, *Pasteurella septica* and *Candida albicans*. This procedure is further developed in the half-plates of blood agar and MacConkey medium with antibiotic discs which give prompt identification and guidance for treatment of pathogens in the urinary tract.

Some drugs have been found to have a special diagnostic significance in certain groups of organisms. Two of these are illustrated : bacitracin for the identification of Group A streptococci, and optochin (ethyl hydrocuprein) for pneumococci.

The transfer of multiple drug resistance is illustrated in **288**.

'Multodisks' made by Oxoid Limited have been used on a number of the plates photographed in this chapter. The following key is provided to the abbreviations printed on the six outer circles which carry the respective drugs:

C	chloramphenicol
E	erythromycin
G	sulphafurazole
P	penicillin
PN	ampicillin
S	streptomycin
TE	tetracycline.

258 *Plate titration of penicillin* On a carpet of *Staphylococcus aureus* were placed porcelain cylinders containing serial twofold dilutions of penicillin, starting from ten units per ml (top left). With the decreasing concentration the size of the zone of inhibition becomes progressively smaller. This method can be used for the titration of antibiotic in a liquid. A curve is plotted relating the diameter of the zone of inhibition to the concentration of the drug. The concentration of an unknown can be read off from this curve once the diameter of the inhibition zone it produces is found, provided a number of experimental variables are adequately controlled. (*Digest agar B, 18 hours at 37°C; reflected light, x¾.*)

259 *Tube titration of streptomycin in small volumes* In the example shown the concentration of the drug was estimated in a sample of patient's serum. A series of twofold dilutions of streptomycin, starting at 64 μg per ml, were made in the bottom row of tubes, and a similar series of patient's serum in the top row. To each tube was added an equal volume of indicator medium which had been inoculated with the Milne streptococcus. Growth of this organism under these conditions produces an acid reaction which clots the medium and changes its colour from red to orange. The bottom row shows that, under the conditions of this test, the minimal inhibitory concentration is 8 μg per ml, the concentration in the fourth tube from the left. In the top row the second tube contains this minimal inhibitory concentration; therefore the undiluted patient's serum in the first tube contains 16 μg per ml. (*The indicator medium contains 1 part horse serum, 1 part 10% glucose, 2 parts saturated solution phenol red; 18 hours at 37°C; reflected light, x1.*)

258

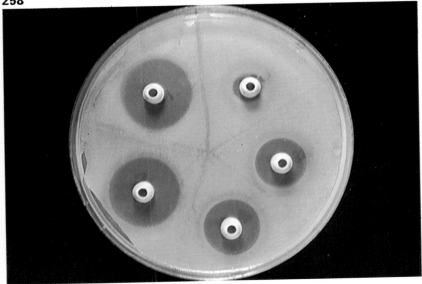

259

149

260 *Mechanisms of penicillin resistance* On a carpet of *Staphylococcus aureus* were placed four discs impregnated with penicillin. One drop of the following was then placed on the respective discs : top left, penicillinase solution ; bottom left, culture of penicillin-resistant staphylococcus ; bottom right, culture of *Streptococcus faecalis* (penicillin-resistant) ; top right, no treatment – control. The result after overnight incubation is shown. Control, top right, disc shows the zone of inhibition of growth by penicillin. This zone is completely eliminated by penicillinase, top left, and by the resistant staphylococcus, bottom left, but not by the resistant streptococcus, bottom right. The resistance of the staphylococcus results from its production of penicillinase ; resistance of *Streptococcus faecalis* does not result from penicillinase production. (*Digest agar B, 18 hours at 37°C; reflected light, x¾.*)

261 *Another result from the same type of experiment* A smaller inoculum of the resistant staphylococcus was placed on the lower left disc. This has reduced the zone of inhibition of growth. There is also a zone of partial inhibition, since it has taken longer to produce the requisite amount of penicillinase from this smaller inoculum of resistant staphylococcus. (*Digest agar B, 18 hours at 37°C; reflected light, x¾.*)

260

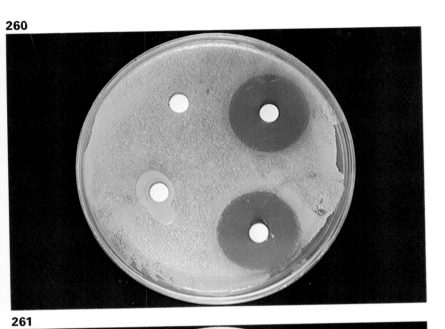

261

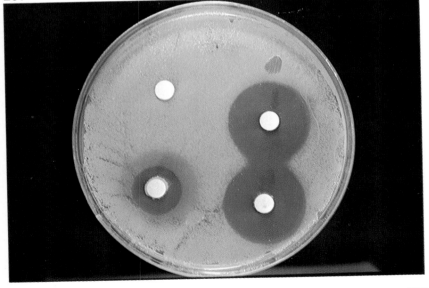

151

262 *Inhibition of sulphonamide by para-aminobenzoic acid*
On a carpet of *Staphylococcus aureus* were placed two red discs containing sulphathiazole; a green disc of para-aminobenzoic acid was placed beside one of them. After incubation it was obvious that the PABA had inhibited the sulphonamide. Culture media which contain PABA are unsuitable for antibiotic sensitivity tests with sulphonamides. (*Oxoid sensitest agar, 18 hours at 37°C; reflected light, x¾.*)

263 *Selection of antibiotic-resistant mutants using a gradient plate* This 15cm diameter petri dish holds a wedge of digest agar overlaid with a second wedge of the same medium containing 0·1 unit of penicillin per ml. The thin edge of the penicillin-containing medium is on the left; thus there is a concentration gradient of penicillin increasing from left to right across the plate. Five strains of *Staphylococcus aureus* were sown in narrow strips along the concentration gradient. After overnight incubation colonies growing at the greatest concentration were re-streaked up the gradient. By repeating this procedure a few times, progressively more resistant mutants were selected. The photograph was taken after three re-streakings, when colonies of four of the five strains were growing at the greatest concentration. (*Penicillin gradient in digest agar, 4 days at 37°C; reflected light, x¾.*)

262

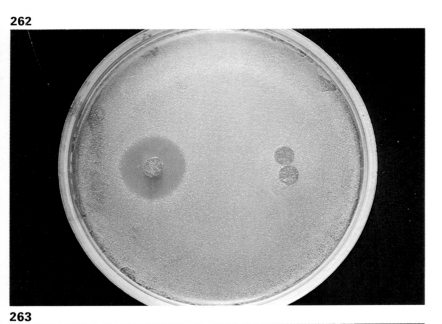

263

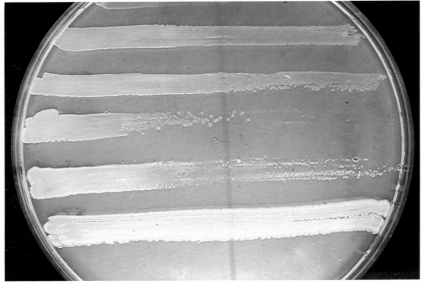

153

264 & 265 *Rapid development of resistance to fucidin* Figure **264** shows zones of inhibition of growth of *Staphylococcus aureus* by fucidin. From this plate inocula were sown on to the plate shown in **265**.

The left half of the plate (**265**) was sown with a loopful of culture from the perimeter (i.e. well away from the zone of inhibition of growth) of the plate shown in **264**. This growth shows the typical sensitivity to the drug shown by organisms in **264**. The inoculum for the right half was taken from the edge of the zone of inhibition ; the result showed that this inoculum contained a resistant variant. Although rapid development of resistance to fucidin can be demonstrated regularly in this type of experiment, the drug is still useful in the treatment of staphylococcal infections. (*Digest agar B, 18 hours at 37°C; reflected light, x$\frac{3}{4}$.*)

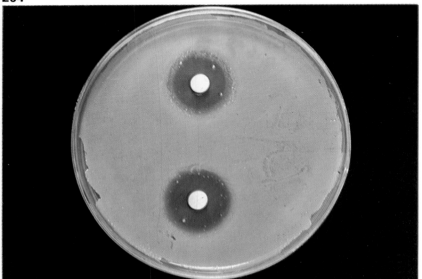

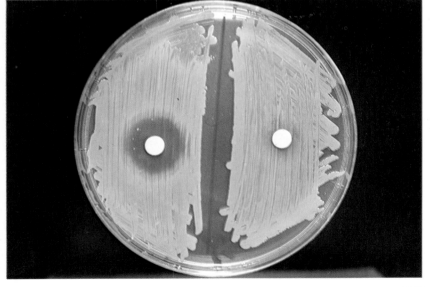

266 *Synergistic action of two antibiotics* On a plate carpeted with *Escherichia coli* were placed two strips of filter paper : the vertical strip contained 8 μg of trimethoprim ; the horizontal strip 250 μg of sulpha-furazole. After incubation a clear zone of inhibition is seen around the trimethoprim strip. This concentration of the sulphonamide has had very little effect on growth, except that it has markedly increased the width of the zone of inhibition by trimethoprim. (*Oxoid sensitest agar, 18 hours at 37°C; reflected light, x$\frac{3}{4}$.*)

267 *Antagonistic action between two antibiotics* Two filter paper strips were placed on a plate which had been carpeted with *Proteus mirabilis*. The white strip (NA) contained 75 μg of naladixic acid ; the orange one (NI) contained 500 μg of nitrofurantoin. Both drugs inhibited growth initially, but the organism later swarmed into the zone of inhibition. The width of the zone around the naladixic acid strip is markedly reduced in the vicinity of the other strip. The zone around the nitrofurantoin is similarly modified by the naladixic acid. (*Digest agar, 18 hours at 37°C; reflected light, x$\frac{3}{4}$.*)

266

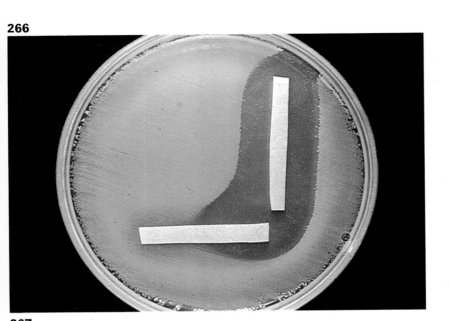

267

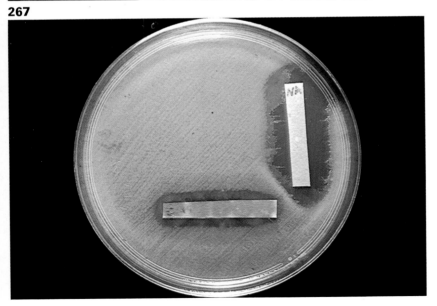

268 *Staphylococcus aureus*, 'practitioner strain' Cultures from the patients of general practitioners are usually sensitive to all of the commonly-used antibiotics. (*Blood agar A, 18 hours at 37°C; reflected light, x½.*)

269 *Staphylococcus aureus*, 'hospital strain' Strains acquired in hospital are usually resistant to many of the commonly-used antibiotics. Compare the diameters of the growth-inhibition zones on this plate with those shown in **268**. Note that a few colonies have grown in the zone of inhibition around the penicillin disc. (*Blood agar A, 18 hours at 37°C; reflected light, x½.*)

270 & 271 *Zones of inhibition of growth around discs containing penicillin* Figure **270** shows an *S. aureus* strain which is fully sensitive to penicillin. The zone of inhibition is wide and the colonies at its margin are minute. The *S. aureus* strain in **271** is producing penicillinase. The zone of inhibition is relatively narrow, and the colonies at its margin are full-sized. (A contaminating proteus is spreading into the top left of the field.) (*Blood agar A, 18 hours at 37°C; reflected light, x6.*)

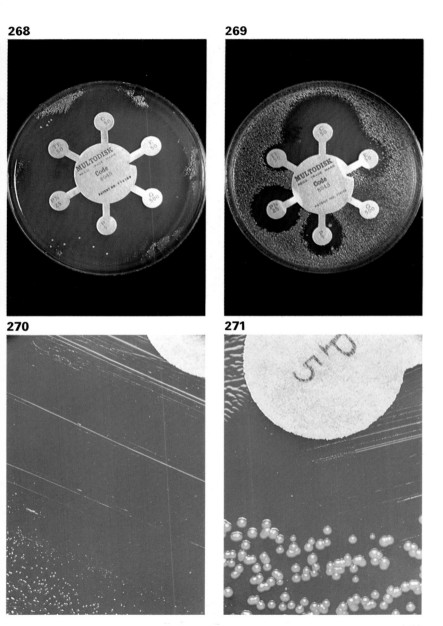

268

269

270

271

159

272 & 273 *Half-plates with antibiotic discs for the identification of urinary tract pathogens* Blood agar was poured into half of the plate, and when this had set MacConkey agar was added to the other half. The plate was sown with a continuous zigzag movement across the two media, and then heavily inoculated in a narrow strip near the junction of the two media. On the strip were placed discs of four antibiotics: nitrofurantoin (NI), naladixic acid (NA), ampicillin (AP) and sulphonamide (SF). This provides a prompt identification of the causative agent of the urinary tract abnormality, as well as suggesting an appropriate treatment.

Figure **272** is *Escherichia coli* infection. This organism grows on both media. In the thin part of the MacConkey wedge it has exhausted the lactose and the pH has reverted to alkalinity. The organism is sensitive to all four antibiotics. (*Reflected + transmitted light.*)

Figure **273** is *Staphylococcus albus* in an acute urinary tract infection. The organism grows satisfactorily only on the blood agar side, with typical opaque white colonies. It is resistant to naladixic acid. (*Reflected light.*)

(*Half-plates of blood agar A and MacConkey agar A, 18 hours at 37°C; x¾.*)

272

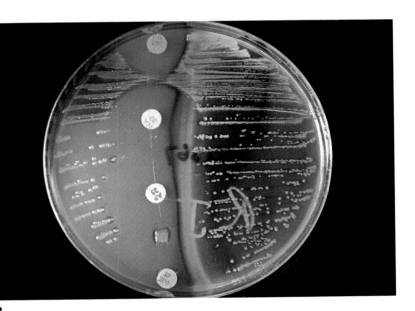

273

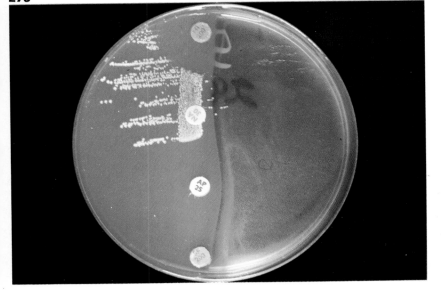

274 *A simple example of how an antibiogram assists in the identification of an organism* *Candida albicans* is resistant to the common antibacterial agents. This together with the characteristic morphology and 'beery' smell of the colonies justifies a presumptive identification of *Candida*. (*Blood agar A, 24 hours at 37°C; reflected light, x¾.*)

275 *Sensitivity of Pasteurella septica to penicillin* *Pasteurella septica* is a somewhat unusual gram-negative rod in that it is quite sensitive to the usual test concentrations of penicillin (the bottom disc labelled P contains 1.5 units). This clue may facilitate recognition of *P. septica*, which is frequently found in wounds resulting from the bites of dogs or cats. (*Blood agar A, 18 hours at 37°C; reflected light, x¾.*)

274

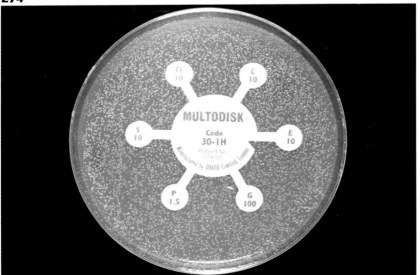

275

163

276 *Bacitracin sensitivity tests* The organism on the left is presumptively identified as a group A streptococcus by its inhibition by bacitracin. The bacitracin resistant streptococcus on the other half of the plate was found to belong to Group G. (*Blood agar A, 18 hours at 37°C; reflected + transmitted light, x¾.*)

277 *Optochin sensitivity of pneumococci* An optochin disc was placed on the blood agar plate which was sown with six test organisms in radial strokes. After overnight incubation optochin has inhibited four stains; these are pneumococci. The other two are alpha-haemolytic streptococci. (*Blood agar B, 18 hours at 37°C; reflected light, x¾.*)

276

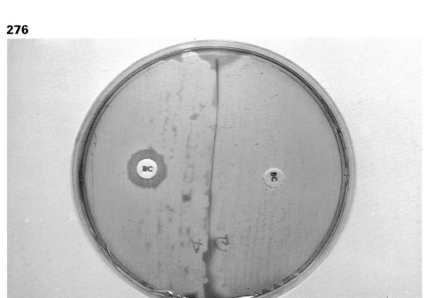

277

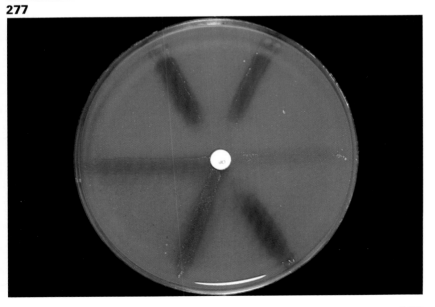

Variation and Genetics

One studies such large populations of microbes that mutations are quite commonplace. These mutants provide useful tools for analysing various microbial functions; they also have provided the basis for many of the modern developments in genetics and in molecular biology. Frequently the microbiologist uses the daughter strain without rigorously proving that it is a true mutant — changes arise from other sources as well as from mutation. For this reason the non-committal term 'variant' is used to denote any progeny that differs from its parent.

Variation may be either phenotypic or genotypic. A genotypic variation represents a change in one of the micro-organism's genes; it may result from mutation or from gene transfer. Phenotypic variation results from a change in the microbe's environment.

It must be emphasised that virtually any property illustrated in this book is subject to variation. Only a few well known examples of phenotypic and genotypic variation are shown in this chapter. Some of the methods for isolating variants are also shown.

Certain variants arise in primary cultures from clinical specimens and one should always be alert to their possible occurrence. Cysteine-dependent coliforms are not uncommon in the urinary tract. *Escherichia coli mutabile* is well known to bacterial geneticists; it also occurs in human faecal samples and, unless it is recognised, it may cause a problem in preparing a pure culture for biochemical testing.

The S → R variation is well known to bacteriologists. Its earmark, a change of colony morphology from smooth to rough, is but one result of the loss of the surface antigen characteristic of the smooth form. Since it is likely to be imperative for diagnosis or epidemiological study that this antigen be identified, bacteriologists are careful to choose smooth colonies for antigenic characterisation.

278 & 279 *Phenotypic variation in cell size of Proteus mirabilis*
The size of the cells of this bacterium varies markedly with the phase of growth. The smear in **278** was prepared from the edge of the swarming growth after 7 hours at 37°C. These very large bacilli are typical of the swarming phase of *P. mirabilis,* and can be shown to have numerous flagella (**173**). The smear in **279** was made at 48 hours when swarming had ceased. The large 'swarmer-cells' have been replaced by smaller cells typical of Enterobacteriaceae. (*Gram, x1,000.*)

280 & 281 *Phenotypic variation in colony size of Group F Streptococci* Streptococci of this group need added carbon dioxide for their full development. The plates shown in these two figures were sown from the same throat swab, from which, as well as the group F streptococcus, larger non-haemolytic colonies also grew. By comparing the number of colonies of this second type in the two figures, it will be appreciated that the area shown in **281** was sown more heavily than that shown in **280**. In the presence of carbon dioxide (**280**) the strepto-cocci have grown into colonies of even size, all surrounded by a wide zone of haemolysis. The plate in **281** was incubated without added CO_2. Even in this heavily inoculated zone, the colonies are much smaller than in the presence of CO_2. In the more sparsely sown areas, the colonies and their haemolytic zones are very small indeed, and many streptococci have failed to grow into visible colonies. (*Blood agar A, 18 hours at 37°C; reflected + transmitted light, x6.*)

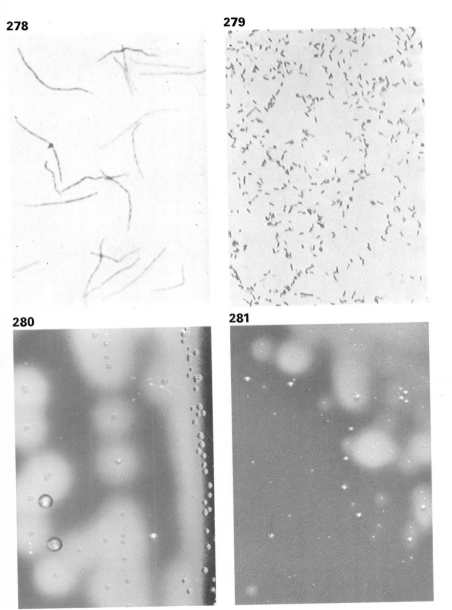

282 *Smooth colonies and rough variants in Escherichia coli*
When it was first isolated this strain of *E. coli* produced smooth colonies with entire edges like the three in the bottom right of the photograph. In serial subculture progressively rougher colonies were isolated. This plate was sown with a mixture of the parent colony and the roughest of its progeny. The rough colonies are also larger, flatter and more opaque than their parents and have a more irregular edge. This type of variation is common in salmonellas and in other Enterobacteriaceae. The change results from the loss of sugars from the polysaccharide of the O antigen. The rough variant differs from the parent form in a number of other respects : it is insusceptible to agglutination or to bacteriolysis with specific antibody, it is relatively avirulent, and because of the lipid on its surface it may spontaneously agglutinate in saline. (*Digest agar B, 18 hours at 37°C; reflected light, x6.*)

283 *Variants which arose in vivo* These colonies grew on a blood plate sown with a throat sample. The colonies without the wide zone of haemolysis are otherwise very similar to the haemolytic ones. Both types were shown to be group G streptococci. This suggests that one is a variant of the other. This could be tested by subculturing each to find whether either would give rise to the other colony type. (*Blood agar A, 18 hours at 37°C; reflected + transmitted light, x6.*)

284 *Cysteine-dependent Escherichia coli from urinary tract infection* Small colonies only were found on this primary culture after overnight incubation. Discs impregnated with cysteine hydrochloride were then applied. Six hours later colonies were fully developed near the disc. Cysteine-dependent coliforms are not uncommon in the urinary tract. (*MacConkey agar A, 24 hours at 37°C; reflected light, x½.*)

285 *Escherichia coli mutabile colonies on MacConkey agar*
The growth is at first the characteristic pale colour of non-lactose fermenters. On these colonies lactose-fermenting mutants arise which produce red papillae. Although this organism is most familiar to geneticists it is occasionally found in faecal specimens. (*MacConkey agar A, 48 hours at 37°C; reflected light, x6.*)

282

283

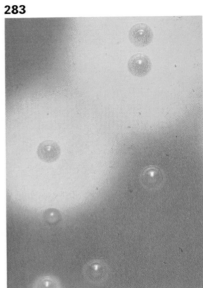

284

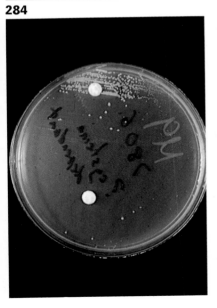

285

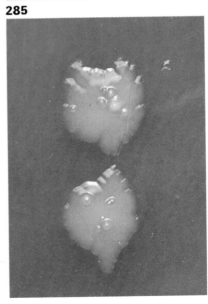

286 & 287 *Selection of auxotrophic mutants* Auxotrophic mutants are defective in their biosynthesis as compared with the prototrophic ('wild type') parent, and will not grow on media which do not contain an essential precursor. After a broth culture of *Escherichia coli* K12 was irradiated to induce mutation, it was treated with penicillin in a minimal medium to kill the parent; parent K12 grows in the minimal medium and is thus susceptible to the lethal action of penicillin, but mutants that do not grow in minimal medium survive as resting cells which are not susceptible to the action of penicillin. Thus this treatment increases the proportion of mutants. The treated mixture was sown on a digest agar plate. When the colonies were just visible the growth was replicated on a minimal agar plate using a velveteen pad. Both cultures were incubated overnight and compared.

In **286** colonies on digest agar, left, which are not represented on minimal agar, are probably auxotrophs. (*Digest agar B and minimal agar; reflected light, $x\frac{3}{5}$.*)

Figure **287** shows identification of the growth requirement of an organism detected by the above method. One of the suspected auxotroph colonies on the digest agar plate was emulsified to form a faintly turbid suspension in distilled water. This suspension was carpeted on three minimal agar plates, on which were placed filter paper discs impregnated with eighteen different amino acids. The organism grew in the vicinity of only one of these discs, the AR shown on this plate. This confirmed that the organism was an arginine-requiring mutant. (*Minimal agar plus discs of arginine, proline, serine, glutamic acid, phenylalanine and tryptophane, 18 hours at 37°C; reflected light, $x\frac{3}{4}$.*)

286

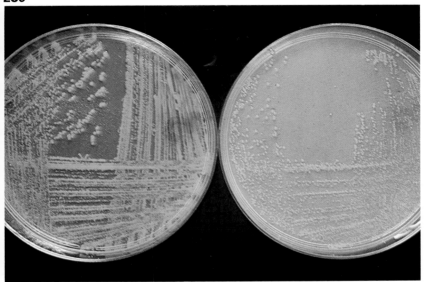

287

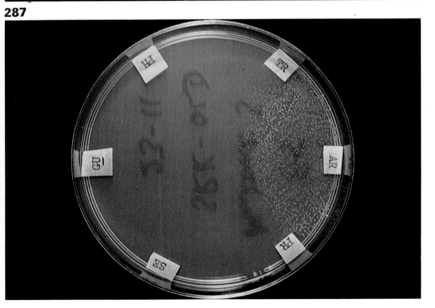

288 Transmission of multiple drug resistance from Salmonella typhimurium to Escherichia coli Strains of *S. typhimurium* (top left) and of *E. coli* (bottom left) were grown together in broth. Note that the *E. coli* parent was sensitive to sulphonamide (G), streptomycin (S), tetracycline (TE), and chloramphenicol (C), while the salmonella parent was resistant to all four of these. From the mixture, *E. coli* progeny (right) which had the multiple antibiotic resistance of the salmonella parent were readily isolated. It can be shown that conjugation is necessary for this transfer of drug resistance. During conjugation an episome is transmitted which consists of two parts : a resistance transfer factor (RTF), and attached genes for resistance to each drug. Multiple drug resistance is widespread among Enterobacteriaceae, and it constitutes a serious problem in the treatment of bacterial enteric disease. (*MacConkey agar, 18 hours at 37°C; reflected light, x ⅔.*)

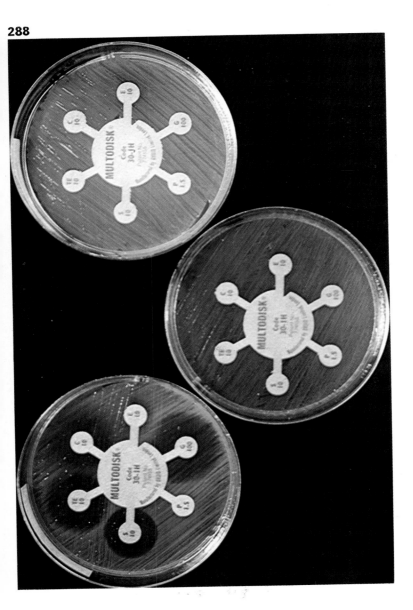

Bacteriophages and Bacteriocins

Bacteriophages are viruses which parasitize bacteria; analogous parasites of fungi are recognized, but they have not been studied so extensively. Phages are of interest as (a) diagnostic tools, as in species identification of *Bacillus anthracis* and *Brucella spp.*, (b) epidemiological tools, as in phage typing of staphylococci and salmonellas, (c) tools for studies in genetics and molecular biology, and (d) vectors of genetic material, as in toxigenic conversion in diphtheria bacilli.

Bacteriophages may be either virulent or temperate for their bacterial host. A virulent phage will cause lysis of the host cells. This may reduce the turbidity of a broth culture, or, in bacteria growing on solid media, it will produce small zones of clearing called plaques. Temperate phages are carried by the host bacteria without causing their lysis; this relationship is termed lysogeny. Sometimes a lysogenic bacterium can be detected by growing it together with a susceptible bacterium when the phage it releases will lyse the second host. In other cases the phage is more firmly integrated into the bacterial genome, in which case more drastic treatment is required to release phage from the lysogenic bacterium. This is done by irradiation of the host bacterium with ultra-violet light before introducing the indicator bacterium, a process known as induction.

Various species of bacteria carry bacteriocins. These have antibiotic properties, and produce areas of inhibition of bacterial growth which may resemble bacteriophage plaques. Bacteriocin typing of some species is useful to the epidemiologist. More specific names, such as colicin, pyocin, diphthericin have been given to bacteriocins of different bacterial species, but since this process seems to be getting a little out of hand, we use the general term bacteriocin.

289 *Clearing of a broth culture by bacteriophage* A culture of *Escherichia coli* in the logarithmic phase of growth was distributed into two tubes. To the one on the left four drops of phage suspension were added ; right, an untreated control. The photograph was taken one hour later. The reduced opacity of the treated tube results from lysis of the bacterial cells. (*Digest broth; indirect transmitted light, x$\frac{3}{4}$.*)

290 *Spontaneous bacteriophage lysis of a culture of a coliform bacterium* Occasionally, routine plate cultures of bacteria reveal evidence of bacteriophage action. One may see colonies from the edge of which a segment has been 'bitten', or there may be plaques on areas of confluent growth. This usually results from a change in the phage from the temperate to the virulent state. Certain phage-bacterium associations seem particularly liable to break down in this way. (*Blood agar A, 18 hours at 37°C; reflected light, x6.*)

291 *Lysogeny in Staphylococcus aureus* Four strains of *S. aureus* were sown in bands ; the same four strains were then spotted onto each band. After incubation any strain which carries a phage which is virulent for one of the other strains is revealed by plaques, by confluent zones of lysis or by a ring of lysis around the spot. (*Digest agar A, 24 hours at 37°C; reflected light, x$\frac{3}{4}$.*)

289

290

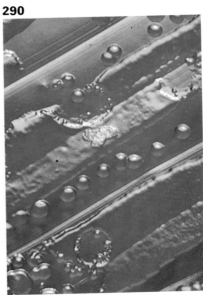

291

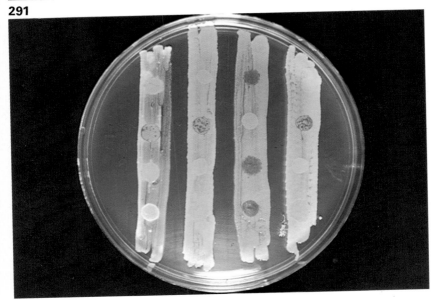

292 *Detection of lysogeny by phage induction* No temperate phage was detected in a strain of *Corynebacterium ulcerans* when it was cross-spotted with other strains (cf. **291**). In a further attempt to detect phage, a thin lawn of the test bacterium was subjected to ultraviolet irradiation to 'induce' any phage it might carry, i.e. to release it from its bacterial association. To detect released phage a series of indicator bacteria were spotted on the irradiated carpet. This view of one spot shows that plaques of two quite different sizes have arisen. In most species of bacteria this would result from the action of two different phages. However, in this case, since subculture of any one plaque gave rise to both types of plaques, the difference in plaque size seemed to result from phenotypic variation in the phage-host association. (*Digest agar, 18 hours at 37°C; indirect transmitted light, x6.*)

293 *Titration of bacteriophage* On a carpet of *Escherichia coli* were spotted tenfold dilutions of phage suspensions from 10^{-1}, top left, to 10^{-9}, bottom right. After overnight incubation the drops of the dilutions 10^{-1} to 10^{-5} have produced confluent lysis. The least concentration to produce confluent lysis, 10^{-5}, is called the routine test dilution (RTD); it, or sometimes 1,000 RTD, is the concentration used in bacteriophage typing. This method may be used also to estimate the number of infective phage particles in the suspension. Six plaques arose from a drop of 0.01 ml at dilution 10^{-7}. Therefore there were 6×10^9 plaque forming units per ml in the original suspension. (*Digest agar, 18 hours at 37°C; reflected light, $x\frac{3}{4}$.*)

292

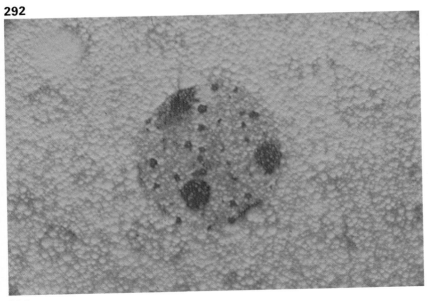

293

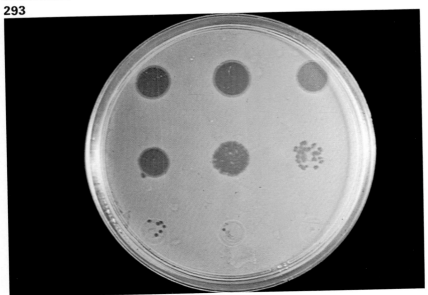

294 *Test of survival of bacteriophage stocks* When stored at 4 °C, most bacteriophages retain their activity for long periods. Before a series of bacteria are phage-typed, the potencies of the stored phages may need checking. The photograph shows how this has been done. Each phage suspension at its routine test dilution has been spotted onto a small rectangular carpet of its host bacterium. Most of the phages have retained their activity; the others would need to be retitrated and/or reprepared. (*Digest agar B, 18 hours at 37 °C; reflected light, x¾.*)

295 *Phage typing of a strain of Staphylococcus aureus* The entire surface of the plate was sown with a broth culture of the staphylococcus. When the inoculum had dried, a series of 24 phages were spotted on the plate. After overnight incubation the phage type of the staphylococcus is determined by the pattern of lysis shown. Phage typing is used in epidemiological studies. (*Digest agar B, 18 hours at 37 °C; reflected light, x¾.*)

294

295

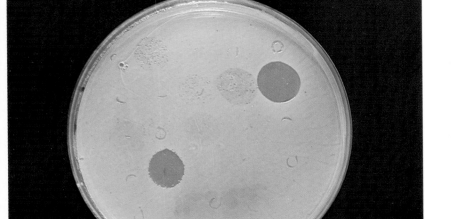

296 *Identification of anthrax bacillus using specific bacteriophage* Four isolates suspected of being *Bacillus anthracis* were carpeted on quadrants of a digest agar plate. A loopful of undiluted phage suspension was placed on each carpet. The growth after incubation shows that the two strains on the right are *B. anthracis*. Since the phage preparation can be stored for a number of years at 4 °C, this is a simple practical method for identifying the anthrax bacillus. (*Digest agar B, 18 hours at 37 °C; reflected light, x¾.*)

297 *Identification of Brucella species using two concentrations of Tbilisi bacteriophage* After a preliminary titration to find the routine test dilution (RTD) of the phage for *B. abortus*, the phage was spotted on to the strain to be identified at two dilutions, 1,000 RTD and 1 RTD. The photograph shows the result with three species of Brucella. *B. suis*, top, is lysed only at 1,000 RTD ; *B. melitensis*, middle, is lysed at neither dilution ; *B. abortus*, below, is lysed at both. (*Serum dextrose agar, 48 hours at 37 °C; reflected light, x¾.*)

296

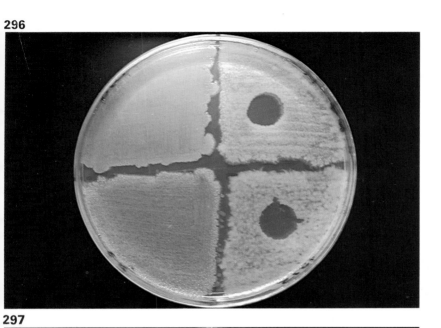

297

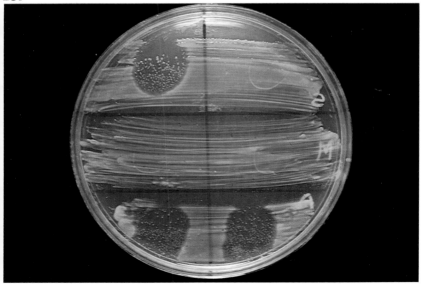

298 *Bacteriocin typing of Pseudomonas aeruginosa* A single line of growth of the test culture was scraped off the plate, which was then exposed to chloroform vapour ; this killed any remaining bacteria, but did not inactivate bacteriocins. Eight typing strains were streaked across the line of the original growth. On further incubation all but one of these typing strains proved to be sensitive to a bacteriocin released from the original culture. (*Tryptose soya blood agar, 1 day at 30°C then 1 day at 37°C; reflected light, x1.*)

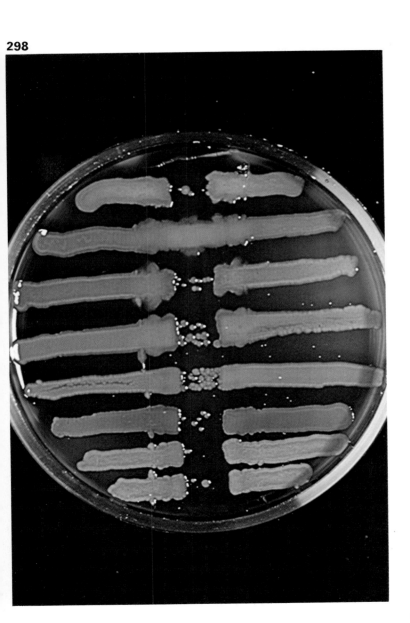

Biochemical Reactions

Many different biochemical tests are used in the identification and classification of bacteria ; fewer are needed for fungi in which morphological features are so much more useful.

In most biochemical tests the colour of the reacting mixture determines the result of the test. Since some of these colours are not readily described, a colour atlas provides scope for illustrating the typical results. Many tests depend for their interpretation on the presence or absence of growth in liquid media ; the test for growth in Koser's citrate medium is shown as representative of these.

It is usually advisable to include positive and negative controls for biochemical tests. With this in mind species appropriate as controls have been chosen for our illustrations.

Tests for biochemical properties of the types illustrated in this chapter are usually applied to selected pure cultures for their identification. In certain special cases the test is applied to primary cultures, as for example in MacConkey medium, where lactose fermentation is used as a key property in the separation of enteric bacteria ; this and other examples will be found in the first section.

299 *Reactions in peptone-water sugars* A variety of carbohydrates may be broken down to acids which change the colour of the pH indicator included in the medium. If gas also is produced, it will be found in the inverted inner tube. From left to right: *Escherichia coli*, acid and gas production; *Pseudomonas aeruginosa*, growth without the production of acid or gas; *Klebsiella aerogenes*, acid and gas production for the first two days followed by reversion of pH to neutrality. (*Glucose peptone water with Andrade's indicator, five days at 37°C; reflected + transmitted light, x¾.*)

299

300 *Fermentation reactions of Corynebacterium diphtheriae gravis in Hiss's serum water sugars* Of the four sugars shown, glucose, sucrose, starch and glycogen, gravis strains ferment all except sucrose (second left). The acidified serum clots and separates out, although it has been remixed in the left hand tube shown. (*Hiss's serum water sugars with Andrade's indicator, 24 hours at 37°C; reflected light, x¾.*)

301 *Reactions in triple sugar iron agar* The medium contains ferrous sulphate as an indicator of hydrogen sulphide production, a pH indicator, glucose (0.1%), lactose and sucrose (each 1.0%). Glucose fermenters produce enough acid to turn the indicator yellow in the relatively anaerobic butt, but not in the slant. Bacteria which ferment lactose or sucrose turn the slant yellow as well. From left to right: unsown control medium ; *Shigella flexneri*, acid butt, unchanged slope ; *Enterobacter aerogenes*, acid butt and slope, abundant gas ; *Proteus mirabilis*, hydrogen sulphide-blackened butt, unchanged slope, gas ; *Pseudomonas aeruginosa*, unchanged butt and slope, no gas. This medium, and others employing similar principles, are extensively used for the preliminary identification of gram-negative enteric pathogens. The medium was kindly donated by Difco. (*Difco triple sugar iron agar, 24 hours at 37°C; reflected + transmitted light, x¾.*)

300

301

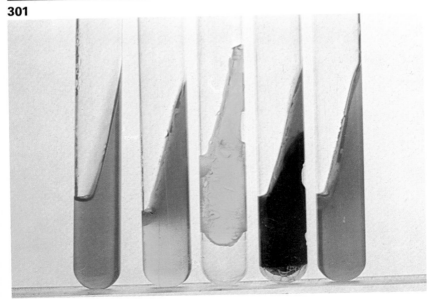

302–305 *Test for oxidation or fermentation of glucose* Two tubes of the appropriate semi-solid medium are sown by stabbing with a straight wire ; one is then sealed to exclude oxygen. Oxidation of the carbohydrate requires oxygen, and will occur only in the unsealed tube. Fermentation does not need oxygen and will occur in both tubes. The indicator is bromthymol blue which is green at neutral pH, yellow at acid, and blue at alkaline pH. The same medium can be used for detecting gas formation or motility.

Pseudomonas aeruginosa (**302**) produces acid only in the open tube, since its breakdown of carbohydrate is oxidative. Note that the acid is formed first in the upper part of the medium, which is near the oxygen. (*18 hours at 37°C.*)

Shigella flexneri (**303**) produces acid in both tubes ; it is fermentative. The acid diffuses throughout the medium and turns it yellow, but the growth is confined to the line of stab-inoculation, since this organism is non-motile. It does not produce gas. (*18 hours at 37°C.*)

Salmonella paratyphi B (**304**) is also fermentative. However, because it is motile, growth is not confined to the line of inoculation but occurs throughout this semi-solid medium. There are bubbles of gas in both bottles. (*18 hours at 37°C.*)

Bordetella bronchiseptica (**305**) grows in the open tube without producing acid. It metabolises the peptone which eventually produces a very alkaline reaction. (*2 weeks at 37°C.*)

(*Hugh and Leifson's medium; reflected + transmitted light, x¾.*)

302

303

304

305

306 *Action of Acinetobacter anitratum on lactose* The two bottles on the left contain 1% lactose peptone water ; the two on the right contain 10% lactose agar slants. Andrade's indicator was used throughout. The two middle tubes were sown with *Acinetobacter anitratum* ; the outer two are unsown controls. After 24 hours incubation *anitratum* has produced acid on the 10% lactose agar slant, but not in the 1% lactose peptone water. The lids should be left loose to enable this characteristic oxidative reaction to develop. (*24 hours at 37°C; reflected + transmitted light, x¾.*)

307 *Test for beta-galactosidase* Some bacteria contain an intracellular beta-galactosidase, yet fail to break down lactose (a beta-galactoside) perhaps for weeks. The enzyme may be released by treating the bacteria with toluene, and it then converts the colourless ortho-nitrophenyl-beta-D-galactopyranoside (ONPG) to the yellow ortho-nitrophenol. The test is especially useful for differentiating late-lactose-fermenters from salmonellae. From left to right : unsown control medium ; *Citrobacter ballerup*, positive ; *Salmonella typhimurium*, negative. (*Saline suspensions treated with toluene at 37°C for 10 minutes, then at 37°C for 30 minutes with added ONPG solution; transmitted light, x¾.*)

308 *Eijkman's test* Production of acid and gas by *Escherichia coli* growing at 44°C (positive reaction) is shown in the centre tube. On the left is an unsown control ; on the right, the negative reaction of *Enterobacter cloacae*. (*MacConkey broth, 24 hours at 44°C; transmitted light, x¾.*)

306

307

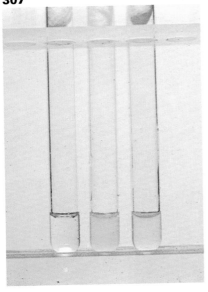

308

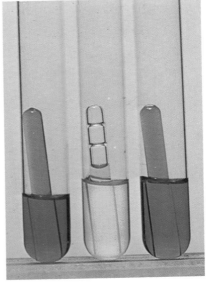

195

309 *Methyl red test* This is a qualitative test of the acidity produced by the growth of a bacterium in phosphate-buffered glucose peptone water. *Escherichia coli* produces a pH of about 5, and hence has a red colour after addition of methyl red. With *Enterobacter aerogenes* the pH never drops so low ; it appears yellow after addition of methyl red. The photograph shows the colours produced after the addition of methyl red in (left to right) : unsown control medium ; *Escherichia coli* culture (positive reaction) ; *Enterobacter aerogenes* culture (negative reaction). (*Glucose phosphate peptone water, 24 hours at 37°C; reflected + transmitted light, x¾.*)

310 *Voges-Proskauer* (*V-P*) *test* From carbohydrate some bacteria produce acetylmethylcarbinol, which in the presence of KOH and air is oxidised to diacetyl. This reacts with alpha-naphthol and a breakdown product of the arginine in the peptone to produce a red colour. The photograph shows the colours produced in (left to right) : unsown control medium ; *Enterobacter aerogenes* culture (positive reaction) ; *Escherichia coli* culture (negative reaction). (*Glucose phosphate peptone water cultures incubated for 48 hours at 37°C; alpha-naphthol and KOH added and shaken for one minute, then allowed to stand for five minutes; reflected + transmitted light, x¾.*)

309

310

311 *Growth in Koser's citrate medium* A test for the ability of an organism to grow in a medium in which citrate is the sole source of carbon. The development of turbidity of the medium is sufficient evidence of growth. It is essential to use a small inoculum, which will not itself render the medium turbid. A positive should be checked by subculture to a second tube. From left to right : unsown control medium ; *Enterobacter aerogenes*, positive ; *Escherichia coli*, negative. (*Koser's citrate medium, 24 hours at 37°C; reflected light, x¾.*)

312 *Test for starch hydrolysis* Growth of a bacterium which hydrolyses starch is recognized by its clear halo when the starch agar plate is flooded with iodine. Above, *Corynebacterium diphtheriae* type *mitis*, negative ; below, *Corynebacterium ulcerans*, positive. (*Starch agar, 2 days at 37°C; 2 minutes after the addition of iodine; transmitted light, x½.*)

313 *Kovac's oxidase test* Oxidase-positive bacteria oxidise phenylene-diamine compounds to indophenol which is purple. Two drops of freshly-made oxidase reagent were placed on filter paper. Above, a colony of *Escherichia coli*, oxidase-negative, was rubbed on to one drop ; below, a colony of *Pseudomonas aeruginosa*, oxidase-positive, was rubbed on to the other. The photograph was taken ten seconds later. (*Cultures grown on digest agar for 18 hours at 37°C; reflected light, x½.*)

311

312

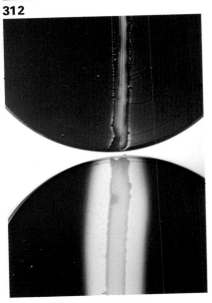

313

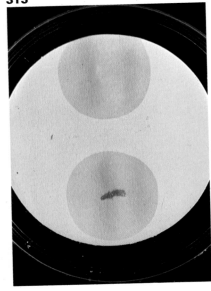

314 *Tube catalase test* Catalase is an enzyme which reduces hydrogen peroxide to water with the evolution of bubbles of oxygen. Since red blood cells also contain a catalase the test should preferably be done on blood-free media. The two outer tubes show the negative reaction of an unsown digest agar slope and the false positive reaction of the blood agar slope. The other two tubes show the positive reaction of *Staphylococcus aureus*, and the negative reaction of *Streptococcus faecalis*. (*Three digest agar B slopes and one blood agar B slope, 24 hours at 37°C; one minute after addition of H_2O_2; reflected light, $x\frac{3}{4}$.*)

315 & 316 *Slide catalase test* One corner of a triangular nichrome wire loop is touched onto the unsown part of a culture plate, then the other corner is touched onto the colony. The loop is immediately placed into a drop of hydrogen peroxide on a slide. Thus one corner carries the test sample and the other provides a control for catalase in the medium.

A positive reaction is seen in **315**. Gas bubbles are produced at the corner of the loop which carried *Staphylococcus aureus*, but not from the other corner which had sampled the uninoculated agar medium.

A negative reaction is seen in **316**. No gas bubbles are produced, either from the *Streptococcus faecalis* colonies which are floating off the top corner of the loop, or from the medium.

(*The reaction of cultures grown on blood agar medium B was photographed by reflected light immediately after immersion of the loop, x4.*)

314

315

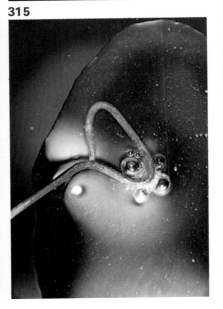

316

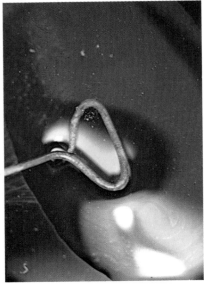

317 *Theory of the nitrate reduction test* All four bottles contained uninoculated nitrate broth. To the bottle second from the left sulphanilic acid and alpha-naphthylamine reagents were added ; there is no colour change because no nitrites are present. To the third tube the same reagents as well as a little powdered zinc were added ; a red colour is developing near the zinc on the bottom of the tube. The fourth tube shows the full colour development about a minute after the addition of the zinc. This reaction sequence will be shown by bacteria which fail to reduce nitrate. Those which reduce nitrate to nitrite produce the red colour without the addition of zinc. Those which reduce nitrite still further produce no red colour even after addition of the zinc. In some cases the nitrite is reduced to nitrogen gas which collects in the inverted inner tube. (*Uninoculated nitrate broth; transmitted light, x¾.*)

318 *A Cook plate for testing reduction of nitrate* The paper strip impregnated with potassium nitrate was applied to the plate ; *E. coli* was stabbed into the right hand side and *Acinetobacter anitratum* into the left. Growth of the nitrate-reducing coli was surrounded by a zone of browned blood agar ; the medium around the anitratum is unaffected. This method is especially suitable for organisms that grow poorly in nitrate broth, but it is liable to contamination. However, contaminants (e.g. the colony at bottom left) are usually readily recognized and cause little confusion. (*Blood agar B, 24 hours at 37°C; reflected light, x¾.*)

317

318

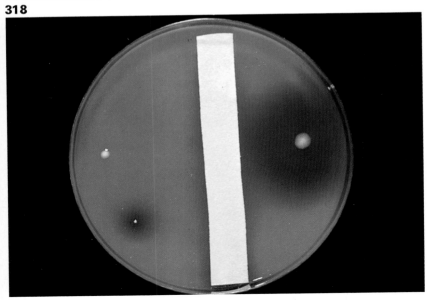

319 *Test for amino-acid decarboxylases of Enterobacter cloacae* These decarboxylases effect the breakdown of amino-acids to amines with the release of carbon dioxide; the resulting increased alkalinity of the medium is detected by the change of colour of a pH indicator. A heavy suspension of the test bacterium was incubated with the amino-acid and bromcresol purple in an appropriate buffer. From left to right: control without amino acid; lysine, negative; ornithine, positive. (*Cowan and Steel's micromethod, 4 hours at 37°C; reflected + transmitted light, x¾.*)

320 & 321 *Tests for indole* Tryptophane, an amino acid supplied by a suitable peptone, is broken down to indole by some bacteria. From left to right: unsown control medium; *Escherichia coli* culture, positive reaction; *Enterobacter cloacae* culture, negative reaction.

Indole (**320**) is volatile and will turn pink a paper saturated with Kohn's reagent. (*Reflected light.*)

A number of methods are available for testing for residual non-volatilised indole in the culture. Kovac's reagent was used in this case (**321**). (*Reflected + transmitted light.*)

(*Peptone water, 24 hours at 37°C; x¾.*)

319

320

321

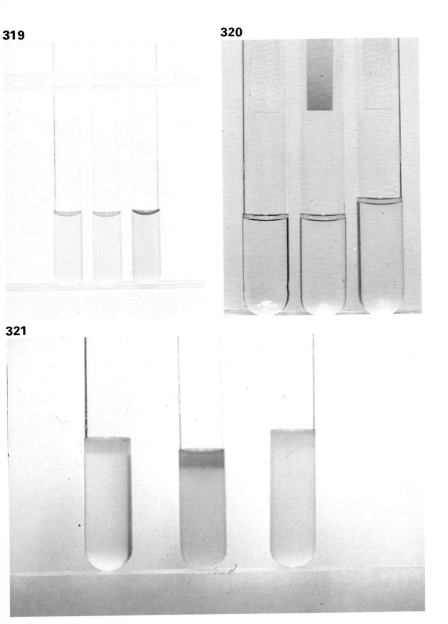

322 *Test for hydrogen sulphide production* Cysteine, an amino acid in digest broth, is broken down to hydrogen sulphide which blackens a lead acetate paper suspended above the culture. This is a very sensitive test : it should be considered positive only if there is obvious blackening after overnight incubation. From left to right : unsown control medium ; *Citrobacter freundii*, positive reaction ; *Shigella sonnei*, negative. (*Digest broth, 18 hours at 37°C; reflected light, x¾.*)

323 *Combined test for motility and hydrogen sulphide production* The semi-solid medium contains (*a*) a tetrazolium salt which enhances the visibility of bacterial growth by staining it red, and (*b*) an iron salt for the detection of hydrogen sulphide. Motile bacteria swim through the medium, but non-motile ones are confined to the line of stab-inoculation. From left to right : *Shigella flexneri*, non-motile, H_2S negative ; *Enterobacter cloacae*, motile, H_2S negative ; *Proteus mirabilis*, motile, H_2S positive. (*Gershman's medium, 24 hours at 37°C; reflected + transmitted light, x¾.*)

322

323

324 *Hydrolysis of gelatin* Gelatin dissolvès in warm water and gels when cool. If it is hydrolysed, it does not gel, but remains liquid when cooled. After incubation these tubes were stood in crushed ice for one hour, and then tilted to find which had gelled. From left to right : *Bacillus anthracis*, negative reaction for hydrolysis of gelatin ; *Bacillus cereus*, positive reaction ; uninoculated control. (*Nutrient gelatin, 2 days at 37°C; reflected light, $x\frac{3}{4}$.*)

325 *Test for liquefaction in gelatin stab culture* If an organism will grow at 22°C, gelatin hydrolysis can be tested by stabbing the gelled medium. The shape of the growing culture and of the zone of liquefied gelatin is characteristic for some species. Left, *Bacillus anthracis*, which has not liquefied the gelatin, but has grown as a series of spikes radiating from the stab – the so-called 'inverted fir tree' ; right, *Bacillus cereus*, which has produced an infundibuliform (funnel-shaped) zone of liquefaction. (*Nutrient gelatin, 3 days at 22°C; indirect transmitted light, x1.*)

326 *Frazier's test for hydrolysis of gelatin* *Pseudomonas aeruginosa* (top) and *Escherichia coli* were grown on gelatin agar. Five minutes after flooding the plate with mercuric chloride the medium appears opaque, except for a zone around the pseudomonas where the gelatin was previously hydrolysed. (*0.4% gelatin in digest agar, 24 hours at 37°C; reflected light, $x\frac{1}{2}$.*)

324

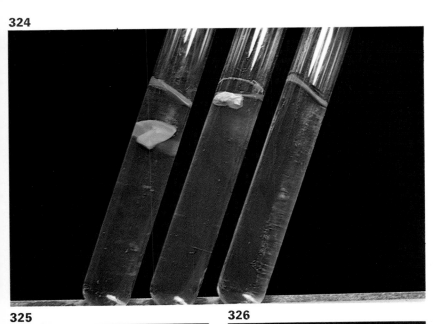

325

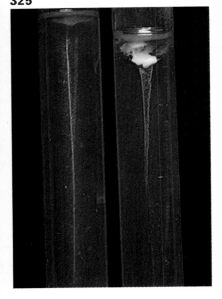

326

327 *Urease reactions on Christensen's medium* Urease-positive bacteria hydrolyse urea to ammonia ; this turns the phenol red indicator a characteristic red-violet colour. From left to right : positive, *Proteus mirabilis* ; negative, *Salmonella typhimurium* ; unsown control. Since the peptone itself may produce an alkaline reaction, a control slope without urea should be included if the reaction develops slowly. (*Christensen's urea medium, 18 hours at 24°C; transmitted light, x¾.*)

328 *Positive urease reaction of Bordetella parapertussis* Heavy suspensions of *B. parapertussis* were made in a urea-containing medium (left) and in base without urea (right). After one hour, the test medium showed the characteristic pink of the positive reaction, while the basal medium was unchanged. This urease test uses non-growing suspensions, and is therefore suitable for the differentiation of *Bordetella spp.* ; *B. pertussis* will not grow on the usual urea medium. (*Lautrop's modification of Ferguson & Hook's method, 1 hour at 37°C; transmitted light, x¾.*)

329 *Breakdown of acetamide by Pseudomonas aeruginosa*
This test has been recommended for the identification of *P. aeruginosa*, especially those strains which do not produce the characteristic pigments. The bottles contain (left to right) : uninoculated control medium ; a simple positive test given by *P. aeruginosa* indicated by a change in the colour of the indicator ; in the next two bottles the red of the positive reaction together with the blue of the pyocyanin produces a deep purple reaction in pyocyanin-producing strains of *P. aeruginosa* ; the tube on the right contains a strain of *P. fluorescens* which has failed to grow at this temperature. (*Acetamide broth, 3 days at 37°C; reflected + transmitted light, x¾.*)

327

328

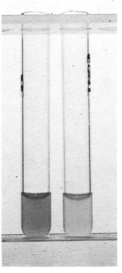

329

211

330 *Test for utilisation of malonate* If a bacterium utilises malonate for growth, the medium becomes alkaline and the indicator turns Prussian blue. With malonate-negative bacteria, the medium remains green, or it may become yellow. From left to right : *Proteus mirabilis*, negative ; *Klebsiella aerogenes*, positive ; uninoculated control. (*Leifson's malonate broth, incubated 24 hours at 37°C with the caps loose; transmitted light, x¾.*)

331 *Test for gluconate oxidation* Some bacteria oxidise gluconate to ketogluconate. The ketogluconate can reduce Benedict's reagent, but the gluconate itself cannot. One ml of gluconate broth culture was mixed with one ml Benedict's reagent, and the mixture placed in a boiling water bath for 10 minutes. From left to right : *Escherichia coli*, negative reaction ; *Klebsiella aerogenes*, positive reaction – an orange-brown precipitate of reduced copper ; uninoculated control. (*Gluconate broth, 3 days at 37°C; reflected + transmitted light, x¾.*)

330

331

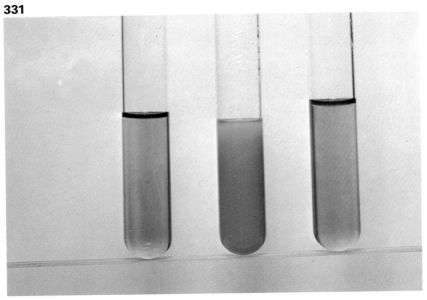

213

332 & 333 *Lecithinase reactions of Bacillus species* Each colony of *B. cereus* in **332** is surrounded by an opalescent zone of precipitated lipid. The lecithinase reaction of *B. anthracis* is usually described as weak (**333**). In the example shown it was negative at 24 hours, although it became more obvious on further incubation. The characteristic curled edges of the anthrax colonies are well shown. (*2.5% egg yolk agar, 24 hours at 37°C; transmitted light, x¾.*)

332

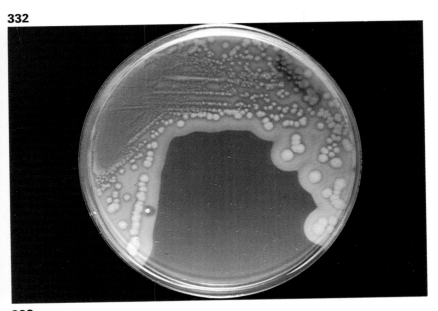

333

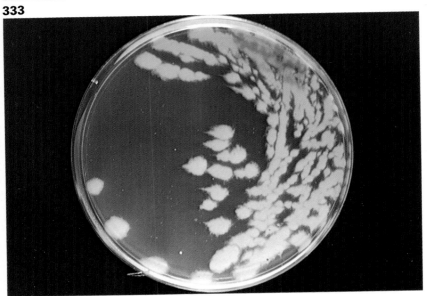

334 *Test for phosphatase* When they grow on phenolphthalein phosphate agar, phosphatase-producing bacteria release free phenolphthalein, and their colonies turn pink when they are exposed to ammonia vapour. Above is *Staphylococcus albus* which is phosphatase-negative; below is *S. aureus* which is phosphatase-positive. (*Phenolphthalein phosphate agar, 18 hours at 37°C; reflected light, x½.*)

335 *Arylsulfatase test* This test is useful in identifying mycobacteria, some of which release free phenolphthalein when they grow in media containing phenolphthalein disulphate. To demonstrate the free phenolphthalein, sodium hydroxide solution is added; this turns the medium pink or red. Left, *Mycobacterium avium*, negative reaction; right, *Mycobacterium intracellulare*, positive reaction. (*Tween albumin broth with 0.001 M tripotassium phenolphthalein disulphate added, 14 days at 37°C; transmitted light, x¾.*)

336 & 337 *Production of levan* Most streptococci of Lancefield group K synthesise a levan from sucrose. On agar containing 5% sucrose their colonies are much enlarged, domed and mucoid.

Figure **336** shows group K streptococcus (syn. *S. hominis*, *S. salivarius*) colonies producing levan. On agar media without sucrose their colonies were much smaller than this.

Group H streptococcus (syn *S. sanguis*) in **337** are levan negative. (*5% sucrose agar, 24 hours at 37°C; reflected light, x6.*)

334

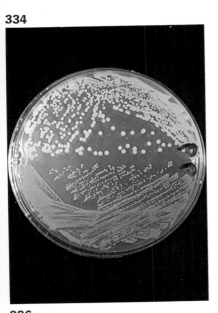

335

336

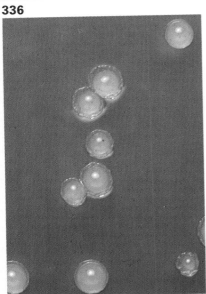

337

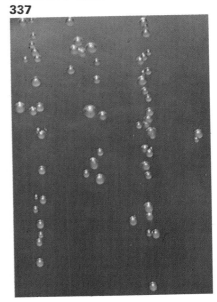

338 *A test for the ability of streptococci to grow on 10% bile agar* The plate was made in two parts – blood agar on one half and bile agar on the other. The test bacteria were streaked across the plate. Some streptococci will grow only on the blood agar, but others grow equally well on both halves. Some streptococci can grow even on 40% bile ; both concentrations are used in routine identification of strepto-cocci. (*Blood agar B + 10% bile agar, 24 hours at 37°C; reflected + transmitted light, x¾.*)

339 *Test for tyrosine decomposition* The plate was prepared with insoluble crystals of tyrosine suspended in digest agar. *Nocardia asteroides* is growing on the left half and *Nocardia brasiliensis* on the right. The photograph was taken through the bottom of the petri dish, hence the bacterial colonies are partly obscured by the crystals. Beneath the area of confluent growth of *Nocardia brasiliensis* the tyrosine crystals have been dissolved. *Nocardia asteroides* has no such effect. (*Digest agar with added tyrosine, 3 weeks at 37°C; reflected light, x¾.*)

338

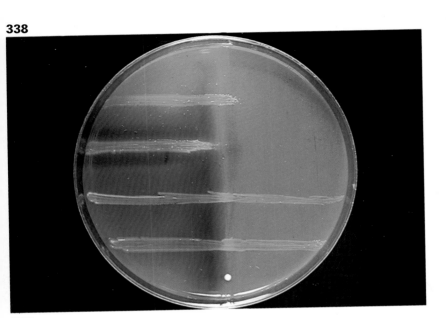

339

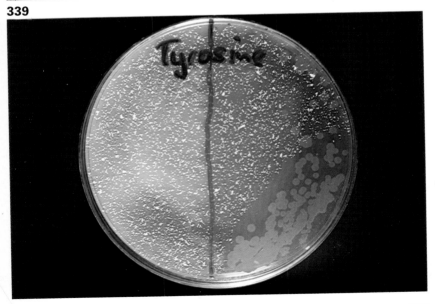

340 *Phenylalanine test* *Proteus* species deaminate phenylalanine to phenyl pyruvic acid which gives a green colour with ferric chloride. From left to right : unsown control ; *Proteus mirabilis*, positive ; *Salmonella typhimurium*, negative. The photograph was taken two minutes after 10% $FeCl_3$ solution was run over the slopes. (*Phenylalanine agar, 18 hours at 37°C; reflected + transmitted light, x$\frac{3}{4}$.*)

341 *Slide coagulase test* Loopfuls of *S. albus* (left) and of *Staphylococcus aureus* (right) were emulsified in water to form thick, even suspensions on a slide. To each a small loopful of undiluted human plasma was added. When the *aureus* preparation was stirred with the loop, within five seconds the cells were clumped into large aggregates which adhered to the slide ; this is a positive reaction. The *albus* preparation shows a negative reaction. A false positive reaction might occur if the plasma contained sufficient antibody ; the resulting agglutination would take longer than 20 seconds to develop, and would not show the tendency to be stuck to the slide by fibrin. (*Cultures grown on digest agar for 18 hours at 37°C; photographed 10 seconds after mixing reagents; indirect transmitted light, x$\frac{3}{4}$.*)

340

341

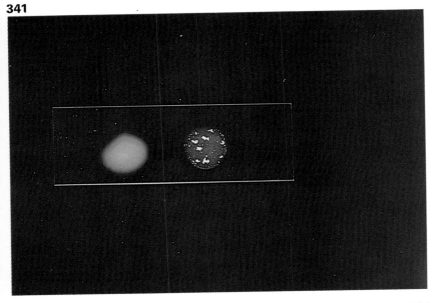

221

342 & 343 *Tube coagulase tests* Five drops of broth culture of *Staphylococcus aureus* were added to 0.5 ml diluted human plasma (**342**), and the mixture incubated for 1 hour. The result is shown in the middle tube – a clot which is obvious when the tube is tilted. The lower tube shows the negative test of a strain of *S. albus*. The upper tube is a negative control in which sterile broth was used instead of broth culture. (*1 hour at 37°C.*) After further incubation (**343**) the clot in the positive tube had contracted away from the wall of the tube. (*2 hours at 37°C.*) (*Reflected light, x1.*)

342

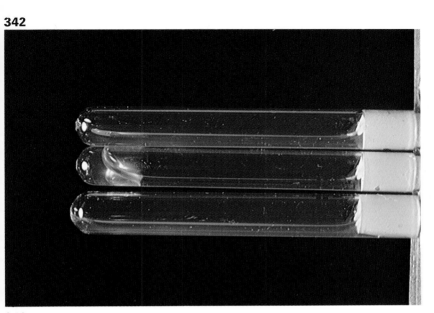

343

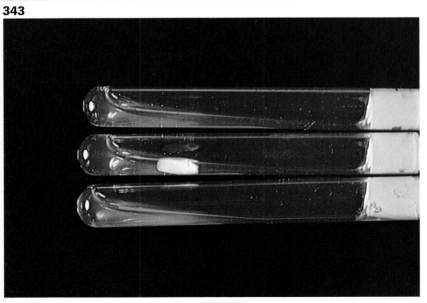

344 *Bile solubility tests for pneumococcus* The two tubes on the left contain broth cultures of pneumococcus; the others contain alpha-streptococci. To one tube of each culture was added 0.1 ml sodium deoxycholate; to the other 0.1 ml distilled water. Within three minutes the pneumococcus was shown to be bile-soluble; the alpha-streptococcus was not. (*Cultures grown in digest broth B, 18 hours at 37°C; indirect transmitted light, x¾.*)

345 *A plate test for demonstration of bile-solubility of pneumococci* With the aid of a hand lens, a loopful of 10% sodium deoxycholate solution was placed on an appropriate area of this primary plate culture from a normal nose. After five minutes re-incubation at 37°C, the pneumococcal colonies in the treated area (below) had lysed. Other colonies in the treated area remain unaffected. (*Blood agar A, 18 hours at 37°C; reflected light, x6.*)

346 *Haemolysin test* Equal volumes of 5% sheep red cells and broth cultures of test organisms were mixed and incubated for two hours at 37°C. The preparations were then centrifuged lightly. Left, *Vibrio el tor* has produced haemolysin; right, *Vibrio cholerae* has not. Tests for lysis of erythrocytes of various animal species are used in identification of many micro-organisms. (*Digest broth, 24 hours at 37°C; reflected + transmitted light, x½.*)

344

345

346

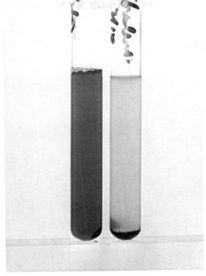

Immunological Reactions

Tests for serum antibodies are used for detecting previous or continuing infection by specific micro-organisms. A titre greater than an agreed level, or the demonstration of a rise in titre over a short period is accepted as diagnostic of certain infections.

Agglutination, precipitation and complement fixation are the serological tests most frequently used. Agglutination is the preferred method if a stable, even suspension of the organism can be made in normal saline; if not, the antigen may be extracted and used in solution in precipitin or complement fixation tests, or it may be attached to the surface of carrier particles as in conditioned haemagglutination tests.

Immunological reactions are used also in the identification and classification of micro-organisms. They are often the most rapid identification methods and, provided adequate controls are used, they are also among the most reliable.

In many instances a species may be subdivided by serological methods into a number of serotypes. In some groups of bacteria, bacteriophage methods provide the most definitive label for any one strain; in still other species, bacteriocin typing may be preferred.

Serological tests may be done in tubes, in the wells of plastic trays, or on plates, tiles or microscope slides; in this chapter a variety of these methods are shown. Each reagent is omitted in turn from successive controls; the concentration of the reagents in controls is the maximum concentration used in the test. This simple method for deciding how a test should be controlled has been followed throughout this chapter. I am aware that the resulting controls are not necessarily those used routinely for some tests. However the method satisfies the object of this book of illustrating principle rather than burdening the student with practical detail.

347 *Identification of a salmonella by slide agglutination* The slide was divided by wax markings into four cells. The photograph was taken half a minute after mixing the re-agents, and shows, top to bottom : negative 'O' (somatic antigen) reaction ; positive 'O' reaction ; negative 'H' (flagella antigen) reaction ; positive 'H' reaction. The wax markings were used to prevent spillage of the large volumes used for photography. *(Indirect transmitted light, x¾.)*

348 *Identification of an organism of the Mycobacterium avium/intracellulare group using tube agglutination* There are three distinct serological types within this group ; these can be recognized by agglutination tests. Mycobacteria form unstable suspensions which readily settle out on incubation. To detect agglutination the organisms are resuspended by lightly tapping the bottom of the tubes. In the negative tube, right, the organisms are easily dispersed on re-suspension. In the positive tube, left, the organisms remain loosely clumped. One should avoid excessive agitation which would break down these clumps. *(Incubated for 18 hours at 37°C; reflected light, x¾.)*

349 *Bacterial tube agglutination titration* In the example shown the test was used in the diagnosis of a case of enteric fever. From left, twofold dilutions of patient's serum starting at 1 in 15 ; right, a saline control. To each tube was added an equal volume of formalinised *Salmonella paratyphi B* suspension. The photograph shows the result after incubation. Large fluffy clumps of agglutinated bacteria have deposited in the first four tubes, leaving clear supernatant fluid. Smaller aggregates are visible as far as the fifth tube ; this represents a titre of 1 in 480 (final dilution after adding suspension). Since the titre of a serum sample taken two weeks earlier was 1 in 30, the rise in titre confirmed the diagnosis of *S. paratyphi B* infection in this patient who had not had TAB vaccine. The test may be used also in identifying an unknown bacterium. The identity of two bacteria is established if (*a*) each agglutinates to titre an antiserum prepared against the other, and (*b*) each absorbs completely the agglutinating antibody from the other's antiserum. *(Incubated 2 hours at 52°C; indirect transmitted light, x¾.)*

347

348

349

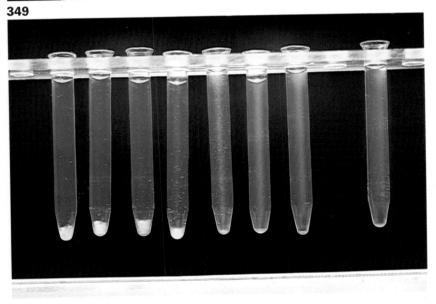

350 *Haemagglutination titration using red cells coated with a bacterial antigen* Sheep red cells which had been treated with 'Old Tuberculin' were used to examine sera for antibodies against tuberculo-polysaccharide. The first and third rows from the top contain twofold dilutions of sera starting at 1 in 10, left. The second and fourth rows contain controls : right, the preceding serum tested against uncoated red cells, and next to it, coated red cells tested with saline only. Agglutinated red cells form a granular film over the bottom of the well ; in the two wells on the left the film has folded on itself. The non-agglutinated cells roll into a small, circular, red button as seen in the wells on the right. Controls are negative, as they should be. Complete haemagglutination is seen as far as the sixth well of the top serum and to the fifth well of the bottom one. These represent titres of 320 and 160 respectively. (*18 hours at room temperature; reflected light, x¾.*)

351 *Titration of four antisera using coated tanned red cells*
Protein antigens can be attached to the surface of sheep red blood cells which have been appropriately treated with tannic acid. Because these tanned cells are somewhat liable to clump spontaneously, controls of tanned cells without attached bacterial protein were included in this test. Row 1 comprises a series of twofold dilutions of an antiserum starting, left, at 1 in 4 ; the well on the right is a control without antiserum. Row 2 comprises the same serum dilution series as row 1. To row 1 were added tanned red cells coated with a bacterial protein, and to row 2, tanned red cells without protein. The agglutination in the first nine tubes of row 1 has detected antibacterial-protein antibody to a titre of 1024. Similarly, rows 3 and 4, 5 and 6, and 7 and 8 are paired titrations of three other sera. Antibacterial-protein antibody is demonstrated in the serum in rows 3 and 5 but not in 7. (*Microtitre tray, incubated 18 hours at room temperature; transmitted light, x¾.*)

350

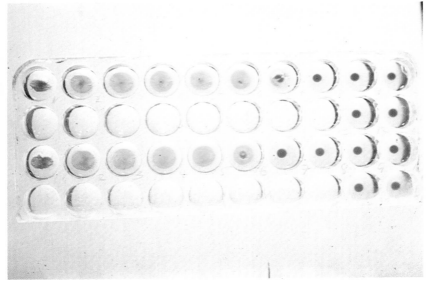

351

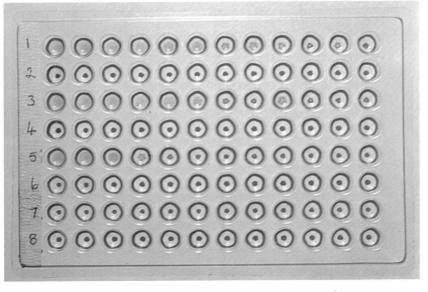

352 & 353 *Agglutination test for the serological diagnosis of Leptospirosis* The antigen, a formalin-killed suspension of *L. ictero-haemorrhagiae*, is liable to spontaneous clumping. Therefore a negative control is included in each batch of tests. The antigen was mixed with serial dilutions of the sera in small volumes. After overnight incubation a drop from each tube was placed on a slide and examined by low power dark field microscopy.

With the negative control (**352**) the leptospires are evenly dispersed throughout the preparation. Because of the depth of the wet preparation, many organisms are out of focus.

With this serum at a dilution of 1 in 1,000 (**353**), the organisms are agglutinated; very few isolated organisms can be seen. With the 1 in 100 dilution of the same serum the clumps were larger. Cross reactions are to be expected with leptospira serotypes. This serum agglutinated *L. canicola* also, but at a lower titre and with a number of loose leptospires. This result confirms a diagnosis of *L. icterohaemorrhagiae* infection.

(18 hours at 4°C; darkfield illumination, x100.)

354 *Specific capsular reaction of a pneumococcus* A drop of peritoneal washings of a mouse infected with a pneumococcus was mixed with a drop of type II pneumococcal antiserum and a drop of Loeffler's methylene blue. Precipitation of antibody globulin on to the capsule dilineates its margin, and the methylene blue stains the bacterial cells, tissue cells and debris. Note that the organisms have been agglutinated into clumps. With pneumococcal sera of other types the capsule could not be seen. By this method a pneumococcus can be promptly typed in pathological fluids such as sputum or C.S.F. *(15 minutes at room temperature; Loeffler's methylene blue; x800.)*

355 *Precipitin test for determining the Lancefield group of a streptococcus* A hydrochloric acid extract of a streptococcus was neutralised and then layered on to streptococcal grouping antisera (A, C, G, from left to right). A zone of precipitation promptly developed in the tube on the left, thereby identifying the streptococcus as group A. The upper and lower menisci of the fluids are clearly shown in each tube, but only the tube on the left has a precipitate at the interface of extract and antiserum. *(5 minutes at room temperature; indirect transmitted light, x¾.)*

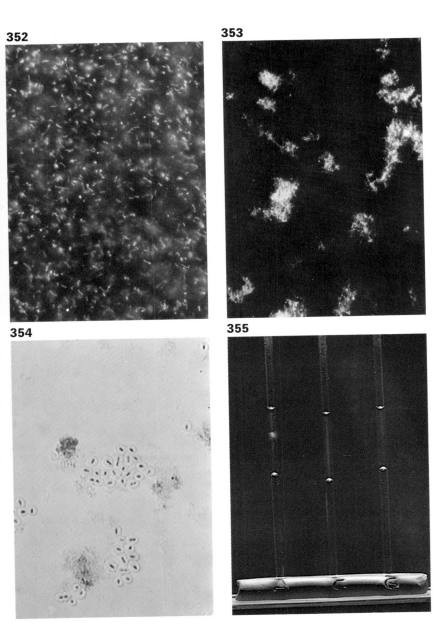

356 *Precipitin reaction, optimal proportions titration* From left, twofold dilutions of antigen, tetanus toxoid, starting from 1 in 10; from right, a control without antigen and a control without antiserum respectively. To each tube was added an equal volume of tetanus anti-serum (1/10); the reagents were mixed well. The tubes were examined at intervals after standing at room temperature. A fine precipitate formed first in the sixth tube from the left; this tube contains the optimal pro-portion of antigen to antibody. (*8 minutes at room temperature; indirect transmitted light, x$\frac{3}{4}$.*)

357 *A hand-lens view of the critical tubes in* **356** The tube which first developed precipitate is third from the left. (*8 minutes at room tem-perature; indirect transmitted light, x1$\frac{1}{2}$.*)

356

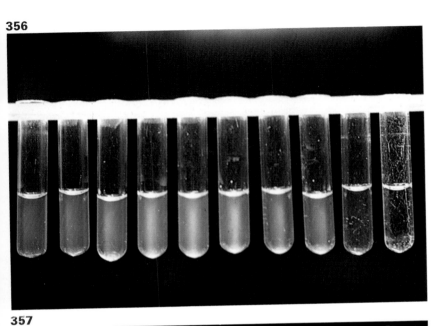

357

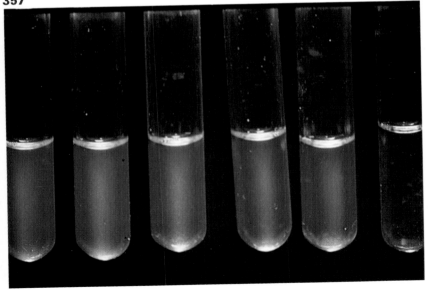

358 *Ouchterlony's method for the comparison of antigens, using the precipitin reaction in agar gel* In one well, bottom, was placed an antiserum prepared against a crude culture filtrate, which was placed in the well at the top. In the left well was placed the culture filtrate of another bacterium, and in the right well a control of uninoculated culture medium. As they diffused towards each other, the reagents formed lines of precipitate at the regions of optimal proportion for each antigen-antibody system. The preparation was then stained for photography. It is clear that the culture filtrate on the top contains at least seven antigens, and the one at the left at least two. None of these antigens were present in the unsown culture medium. The identity of the antigens in the two preparations is established by their lines joining ; lines which cross each other belong to non-identical systems. (*Amido Schwarz stain; transmitted light, x1* $\frac{1}{2}$.)

359 *Test for antibodies to Aspergillus fumigatus in a patient's serum* The patient's serum was placed in the centre well (P) and control anti-fumigatus serum (S), prepared in a sheep, in the left and right wells. Two different extracts (1 & 2) of *Aspergillus fumigatus* were used ; each was placed in a large and a small well to vary the amount of antigen. It is clear that the patient has antibodies against the aspergillus, and that some of them are homologous with those in the sheep serum, while others are not. Note too, that while the extracts 1 and 2 contain homologous antigens as shown by the reaction of identity of lines in the areas between them, extract 2 has detected more antibodies than extract 1 has. (*Amido Schwarz stain; transmitted light, x1.*)

358

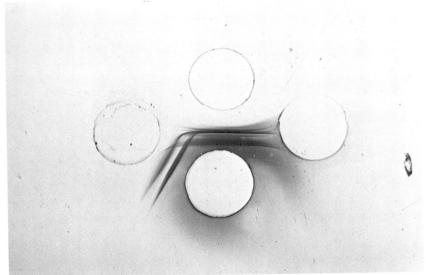

359

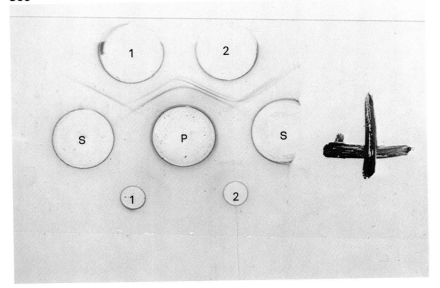

360 *Gel-diffusion plate to demonstrate toxigenicity of diph-theria bacilli* In the medium, which favours the production of diph-theria toxin, was submerged a filter paper strip saturated with antitoxin. Three strains were then sown to produce lines of growth at right angles to the paper strip. Diphtheria-toxigenic strains developed V-shaped lines of precipitate in the zone of optimal antigen-antibody proportion. It is clear that the middle strain and the one on the left are producing the same toxin, since their precipitin lines merge (reaction of identity). This is a convenient *in vitro* method for identifying a toxigenic strain of diphtheria bacillus. (*Elek's medium, 2 days at 37°C, then 2 days at 4°C; transmitted light, x$\frac{1}{2}$.*)

361 *Gel diffusion plate to demonstrate epsilon toxin-producing types of Clostridium welchii* A trough which was cut in a serum agar plate was refilled with agar containing commercial 'pulpy-kidney' antiserum. This serum was prepared against *C. welchii* type D, and con-tains predominantly epsilon antitoxin. Across the plate were streaked three strains of *C. welchii* : left, type A ; middle, type B ; right, type D. The dense precipitin line shows that epsilon toxin is produced by types B and D (reaction of identity) but not by type A. In the same way an unknown strain could be identified as an epsilon toxin-producer, and therefore a potential cause of enterotoxaemia of animals or occasionally of man. The other lines represent other soluble antigens of the clostridia, including other exotoxins. (*Serum agar, 2 days at 37°C, then 2 weeks at 4°C; darkfield illumination, x$\frac{1}{2}$.*)

360

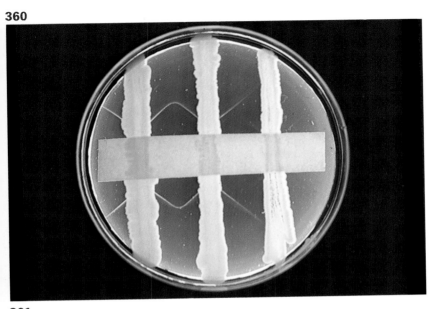

361

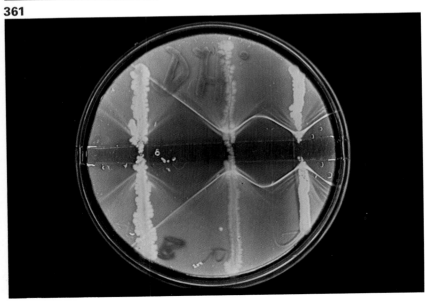

362 *Immunoelectrophoresis of a bovine serum sample, above, and of its albumin fraction, below* Complex antigens can be compared by a combination of electrophoresis and precipitin reaction in a gel. Each sample was placed in a small circular well in a block of agar on a glass slide. A current was passed through the gel to separate the antigens by electrophoresis. When the current was discontinued, antiserum was placed in the long trough. As the reagents (antibodies in the trough and antigens distributed throughout the gel) diffused towards each other, precipitin lines developed (cf. **358**). Two methods are now available for comparing antigens : (i) their shape and position in the gel, and (ii) reactions of identity, partial identity or non-identity as revealed by joining, spurring or crossing of their arcs respectively. The arcs near the cathode (–) end of the pattern represent various globulins. The location of the typical albumin curve at the anode (+) end is clearly shown. Most of the non-albumin components have been removed from the albumin fraction, but some contaminating globulin has been detected by this test. (*Amido Schwarz stain; transmitted light, x1 ½.*)

363 *Immunoelectrophoresis in the diagnosis of farmer's lung* *Micropolyspora faeni* extract was placed in the two origins marked M, and *Thermoactinomyces vulgaris* extract in that marked T. After their electrophoresis two patients' sera were placed in the troughs. The lines which have developed demonstrate that the serum in the upper trough contains a number of antibodies against different components of *M. faeni*, while that in the lower trough contains at least antibodies against *T. vulgaris* detected by this test. The test has demonstrated antibodies in each serum against only one of the two fungal extracts. (*Amido Schwarz stain; reflected light, x1 ½.*)

362

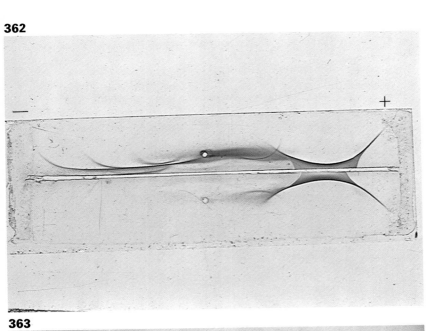

363

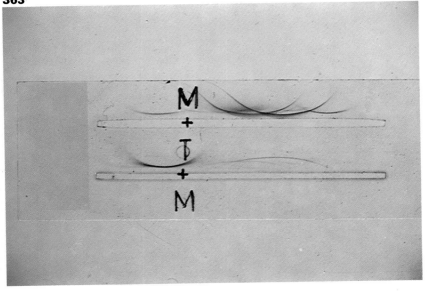

241

364 & 365 *Action of a commercial tetanus antitoxin on a blood agar plate culture of Clostridium tetani* Haemolysis by *C. tetani* and its inhibition on half of the plate by antitoxin is seen in **364**. Since the toxin is distinct from the haemolysin, the antiserum must have contained anti-haemolysin as well as antitoxin. ($x\frac{3}{4}$.)

Close-up of the border between the two zones (**365**) shows that colony morphology also is changed by the commercial antitoxin ; the antiserum also contained antibody which inhibited the development of the characteristic spreading colony. (*x6.*)

(*Blood agar B, 24 hours at 37°C; reflected light.*)

364

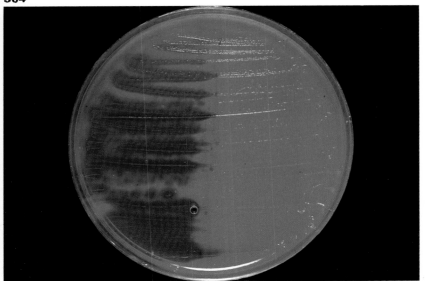

365

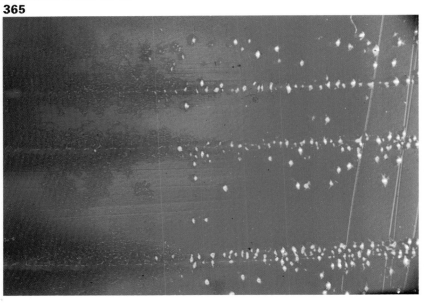

366 *Titration of anti-streptolysin O in a patient's serum* From left to right the tubes contain 1 ml volumes of serum diluted 1 in 100, 200, 333, 500 and 1,000. To each tube was added equal volumes containing 1 unit of reduced streptolysin O and washed horse red cells. On incubation lysis of the red cells has been inhibited in the first four tubes; hence the titre of the patient's serum is 1 in 500, i.e. the serum contains 500 units of anti-streptolysin O per ml. Late complications of streptococcal infection are usually accompanied by anti-streptolysin O levels above 200 units per ml. Appropriate controls for this test are shown in **367**. (*1 hour at 37°C; transmitted light, x¾.*)

367 *Controls for the anti-streptolysin O titration* Twofold dilutions of a known antiserum were examined under the same conditions as the test sera. That the control serum inhibited lysis in the second tube, which contained 1 unit anti-streptolysin O, but not in the third, which contained ½ unit, showed that the sensitivity of the test was correct. (*1 hour at 37°C; transmitted light, x¾.*)

366

367

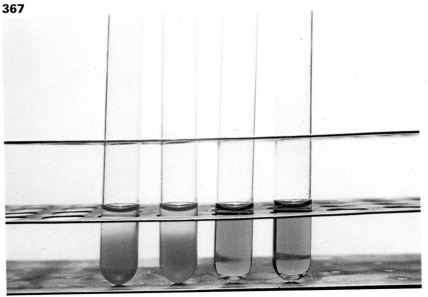

368 *Nagler reaction for the identification of Clostridium welchii* The egg yolk agar plate was sown with a culture of *C. welchii* Type A. On one half of the plate each colony is surrounded by an opalescent zone of precipitated lipid released from lecithin in the egg yolk by lecithinase. These zones are absent from the other half of the plate which had been treated previously with *C. welchii* antitoxin. This provides a specific identification of *C. welchii*, provided *C. bifermentans*, which produces an antigenically similar lecithinase, is excluded on other characters. (*5% egg yolk agar with 5% Fildes' peptic digest of blood, 18 hours anaerobically at 37°C; darkfield illumination, x¾.*)

369 *The use of growth-inhibition by antibody in the identification of a mycoplasma* The dark semicircular area (right) is a filter paper disc which was saturated with antibody, and applied to a plate which had been sown previously with a mycoplasma. The antibody has inhibited growth in the vicinity of the disc, thereby confirming that the mycoplasma is of the same species as that against which the antibody was prepared. (Note that bacterial growth is not inhibited by specific antibody, e.g. colonies have grown in the presence of antiserum in **368**.) (*Whittlestone's medium, 1 week at 37°C; unstained, x10.*)

368

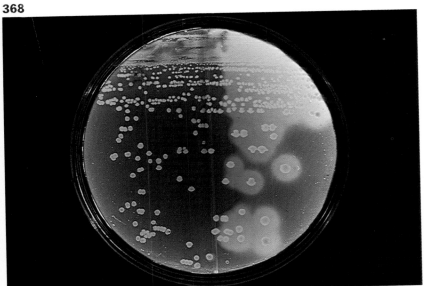

369

370 *Titration of haemolytic antibody for use in the complement fixation test* The antibody was prepared in a rabbit against sheep red cells. From left, twofold dilutions of antibody starting from 1 in 100 ; right, a control containing saline instead of antibody. To each tube was added one volume of washed 2% suspension of sheep red cells, one volume of excess complement and two volumes of saline. The photograph shows the result after incubation. Lysis of red cells has occurred in the first three tubes. This represents a titre of 1 in 400 ; the third tube is said to contain 1 minimal haemolytic dose (MHD). In the complement fixation test 3 MHD will be used ; that is, the haemolytic antibody will be used at a dilution of 1 in 133. The amount of complement to be used in the CF test is titrated in the same way, using 3 MHD of antibody. The preliminary titration of haemolytic antibody and of complement must be carried out under the same conditions as the CF test itself ; the two volumes of saline in the preliminary titrations take the place of the test serum and antigen in the CF test. (*Incubated 30 minutes at 37°C; indirect transmitted light, x$\frac{3}{4}$.*)

371 *Titration of complement fixing antibody to influenza virus* From left, the first six tubes contain twofold dilutions of patient's serum starting at 1 in 100 ; then four controls : (*a*) without patient's serum, (*b*) without haemolytic antiserum, (*c*) without complement and (*d*) without influenza antigen. To each tube, except the appropriate controls, was added one volume of influenza antigen and one of complement (3 MHD). After preliminary incubation, one volume of sheep red cells and one of haemolytic antibody (3 MHD) were added. (All tubes contained the same final volume ; the differences in height of the liquid resulted from differences in bore of the tubes.) The photograph shows the result after reincubation. Note first that all four controls have given the expected result : they show that neither patient's serum alone nor influenza virus alone fix complement, and that both haemolytic antiserum and complement are necessary for lysis. Complement was fixed by the influenza antigen-antibody mixture in the first two tubes, as shown by their non-lysis ; hence the titre of complement-fixing antibody in the patient's serum was 1 in 200. (*Both primary and secondary incubations were for 30 minutes at 37°C; indirect transmitted light, x$\frac{3}{4}$.*)

370

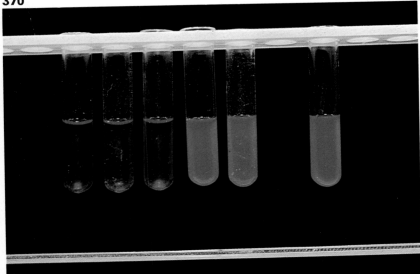

371

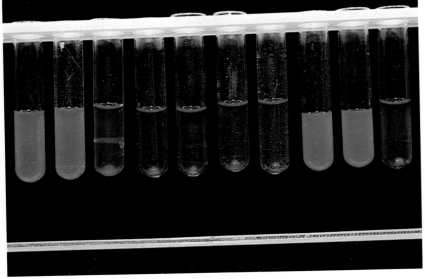

372 *Bactericidal action of antibody plus complement* One drop
of overnight peptone water culture of *Salmonella pullorum* was distributed to each well in row A, one drop of the culture diluted tenfold
to each well in row B; rows C, D, E, F, and G are further tenfold dilutions
to 10^{-6} in row G. Then, one drop of antibody was distributed as follows:
at dilution 1/50 to each well in column 1, at 1/100 to each well in
column 2, ... up to 1/100,000 to each well in column 8. Column 9 was
a control series without complement; column 10 was a control series
without antibody. After one hour's incubation one drop of complement
(1 in 10 dilution) was added to each well (except of course those in
column 9). After a further hour's incubation, growth medium containing
tetrazolium chloride was added; the preparation was then incubated
overnight and examined. Surviving bacteria have grown and produced
a red button stained by the reduced tetrazolium chloride; the bactericidal
effect of antibody acting with complement is expressed by absence of
growth. Note first the prozone, the absence of killing in column 1, presumably resulting from deviation of complement by excess antibody.
Second, in well 7G an estimated 15 cells (see **374**) were killed by antibody at an original dilution of 1 in 50,000; in well 7F the same dilution
of antibody has killed 150 bacteria. In wells 3C and 4C, an estimated
150,000 bacteria were killed; in these wells the proportion of antibody
to antigen was near the optimum for bactericidal activity. Third, in the
first and ninth columns the antibody is sufficiently concentrated to produce an agglutination pattern somewhat similar to that seen in **350**.
(*Transmitted light, x$\frac{1}{2}$.*)

**373 *Same titration as 372, half the concentration of comple-
ment*** This titration was put up at the same time as the one above, but
the concentration of complement was half that used for the previous
figure. Note that this has produced the following results: (i) the bactericidal effect is confined to fewer wells; (ii) the prozone is more
marked; (iii) even at this dilution of complement, in well 7G an estimated 15 cells have been killed by antibody at an original dilution of 1
in 50 `00. These results illustrate the remarkable potency of antibody
and complement in killing gram negative pathogens *in vitro*. (*Transmitted light, x$\frac{1}{2}$.*)

372

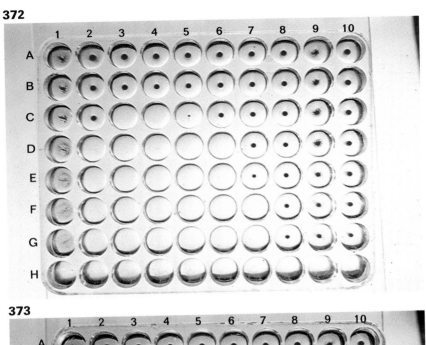

373

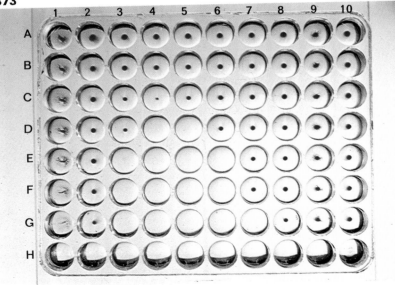

374 *A viable count of the salmonella used in* **372 & 373** Six drops, each of 0.02ml of the 10^{-6} dilution, were placed on the surface of this MacConkey agar plate. After they had dried the plate was incubated and the colonies counted. In all there were 93 colonies, an average of 15.5 (say 15) colonies per drop. With this data the viable count of the original suspension could be estimated as $15 \times 50 \times 10^6 = 7.5 \times 10^8$ viable units per ml. It had been found previously that counts on MacConkey and blood agars were not significantly different. (*MacConkey agar, 18 hours at 37°C; reflected light, x¾.*)

375 *Selection of a particular flagella antigen phase of a diphasic Salmonella, using the Craigie tube* The salmonella was inoculated on to the semi-solid agar inside the top of the inner tube, which is open at both ends. Some hours later the organisms are seen to be migrating down the inner tube from which they will escape to the outer part. Organisms of the known phase are immobilised by added antiserum. Therefore, with appropriate conditions, organisms of the alternative, unknown phase can be grown from the outer tube. (*Semi-solid digest agar, 6 hours at 37°C; indirect transmitted light, x¾.*)

374

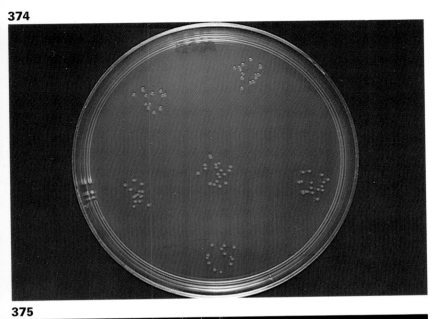

375

376 *Detection of syphilitic antibody in a patient's serum using fluorescent antiglobulin* A smear of *Treponema pallidum* was exposed to the serum of a patient with syphilis. Later, the excess serum was washed off, and the preparation was similarly treated with fluorescein-conjugated anti-human-globulin. The smear was then examined by ultraviolet microscopy. The spirochaetes can be seen in this positive reaction because the antibody was bound to them and to the fluorescent antiglobulin. Since human serum may contain cross-reacting antibody against commensal treponemes, the patient's serum was first absorbed with an extract of the Reiter treponeme. The test is therefore called the fluorescent treponemal antibody absorption (FTA-ABS) test. *(Wellcome reagents, x1000.)*

377 *A positive Heaf test for hypersensitivity to tuberculin* A dose of tuberculoprotein was pricked into the skin with a six-tined applicator. After three days, the site is raised into a firm plaque. Vesicles which had formed at the sites of injection have ruptured, and necrosis is beginning in the central zone. There is a wide diffuse area of hyperaemia, but this is not obvious in the photograph. This would be classified as a grade 4 reaction. *(Reflected light, x$\frac{3}{4}$.)*

376

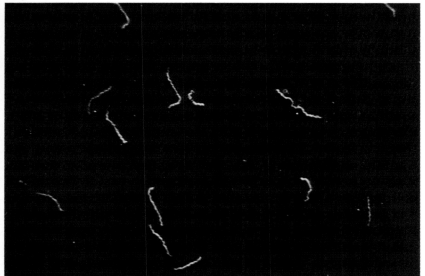

377

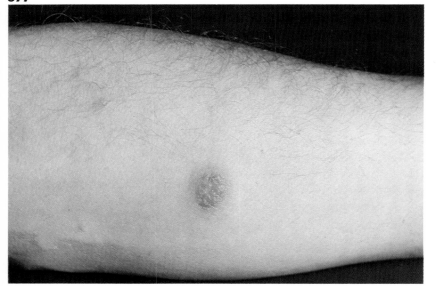

Pathogenicity Tests

Laboratory animals are extensively used in the experimental study of infectious disease and in the assay of chemotherapeutic agents and immunological preparations.

In the identification of a pathogen a laboratory animal may be used at either of two stages of the investigation. First, it may be inoculated directly with material from the patient. Here, as in the isolation of pneumococci or tubercle bacilli, one uses the special sensitivity of the animal to the organism or its particular ability to select the pathogen from a mixture of organisms in the inoculum. Second, one may infect the animal with a pure culture in an attempt to produce a typical disease in the animal to confirm the identity of the organism, or to verify its virulence ; with some of the pathogenic fungi characteristic morphological structures are produced only *in vivo*.

Animal pathogenicity tests are frequently controlled by the use of specific neutralising antisera. This provides prompt specific identification of a pathogenic organism or its toxin.

A number of our dissections have been illuminated from the caudal end ; this proved to be the most satisfactory method of preventing the smaller abdominal organs being obscured by the shadow of the liver. The photographs have been printed with the head uppermost ; this is the simplest orientation for the viewer, although the lighting, which now appears to come from below, may appear unnatural to some.

378 *Corynebacterium diphtheriae – intradermal test for toxigenicity* Suspensions of a number of strains were injected into the skin of this rabbit. Three strains have produced a white zone of necrosis surrounded by erythema. That lesions failed to develop in an antitoxin-protected rabbit confirms that these are caused by diphtheria toxin. Similar lesions can be produced in the guinea-pig. (*48 hours after infection; reflected light, x$\frac{1}{2}$.*)

379 *Skin lesions in a guinea-pig resulting from infection with Corynebacterium ulcerans* A greater dose might have caused an acute toxic death, which in some strains may have been inhibited by *C. diphtheriae* anti-serum. This ulcer resulted from sloughing of necrotic skin three days after infection. (*Photographed six days after infection; reflected light, x1.*)

378

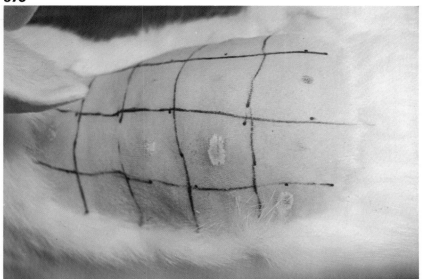

379

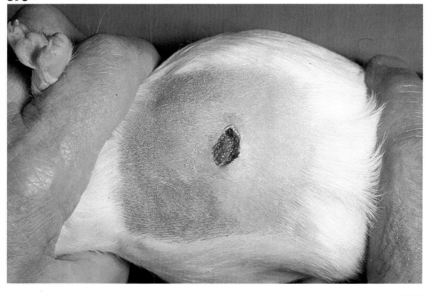

380 *Local subcutaneous reaction to Corynebacterium diph-theriae in a guinea-pig* The animal had received 0.2ml of turbid sus-pension of a Loeffler's slope culture of a *gravis* strain. A pale, necrotic area *(arrow)* is surrounded by an oedematous zone with dilated blood vessels. (*48 hours after infection; reflected light, x¾.*)

381 *A deeper dissection of the guinea-pig shown in* 381 The adrenal has assumed a colour like that of the nearby liver; compare it with the normal adrenal shown in **382**. This is characteristic of diph-theria. The specificity of these lesions can be checked by protecting a control guinea-pig with diphtheria antitoxin. (*48 hours after infection; reflected light, x¾.*)

382 *Normal guinea-pig adrenal* A normal guinea-pig adrenal has a bright orange-yellow colour. Compare this with that shown in **381**. (*Reflected light, x¾.*)

380

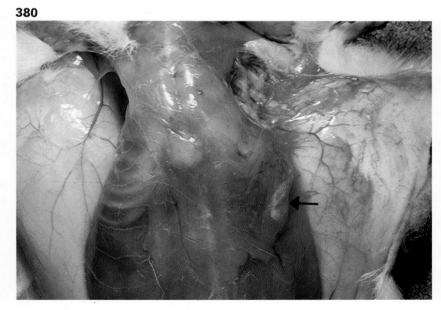

381

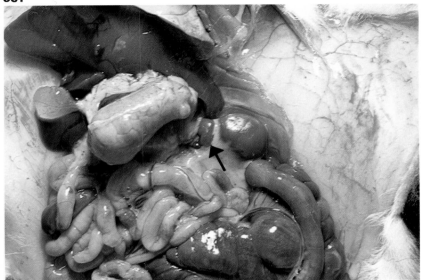

382

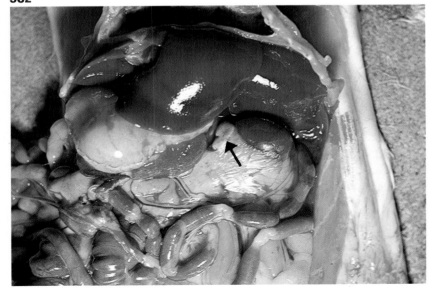

383 *Tuberculosis in a guinea-pig* This animal died six weeks after injection of 0.01mg human tubercle into the muscles of *its* right thigh (*arrow*). Note the chain of swollen lymph nodes, including the regional (precrural) lymph node which is near the left edge of the picture. There are tubercles in the liver and lungs and, although they are less obvious in this picture, in the spleen. Kidney lesions are absent. The caseous lesion at the site of inoculation has been dissected. The bovine tubercle bacillus produces a similar picture to this. The avian type and most other mycobacteria fail to produce such a generalised infection. (*Reflected light, x1.*)

383

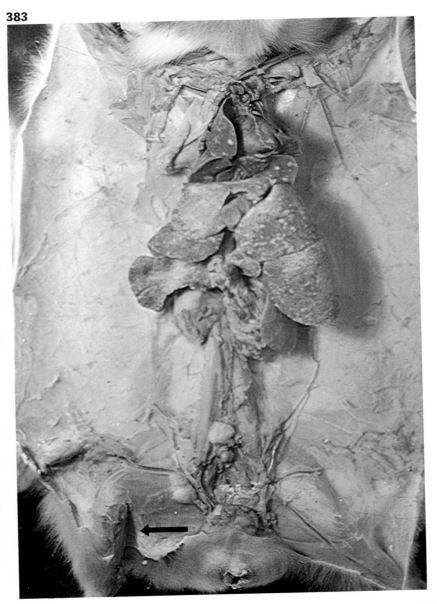

263

384 & 385 *Tuberculosis in a rabbit* This followed intravenous injection of 0.04mg of mammalian tubercle bacilli. Formalin-preserved museum specimens. (*Reflected light, x¾.*)

Figure **384** shows infection with bacilli of the bovine type. This animal died at four weeks with extensive pulmonary involvement; when the chest cavity was opened the lungs remained distended by tuberculous tissue. This rabbit also had many lesions in the spleen, kidneys and mesentery.

Figure **385** shows infection with bacilli of the human type. There are fewer lesions (cf. **384**), and most of them are smaller and have an ill-defined edge. There is plenty of relatively normal lung tissue between the lesions, so that, when the thorax was opened, the lungs collapsed — note the space between the dorsal edge of the lung and the ribs.

384

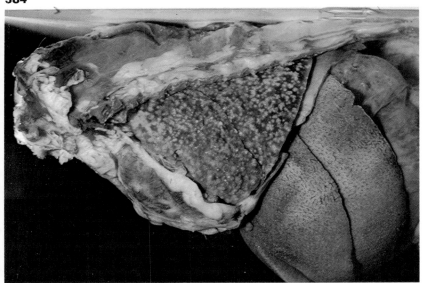

385

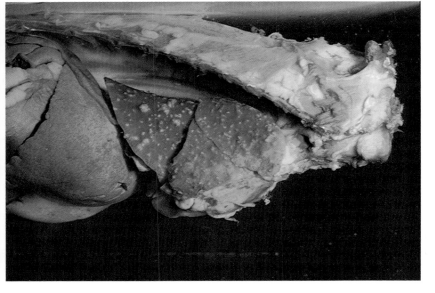

386 *Experimental pseudotuberculosis in a guinea-pig* Following
intramuscular inoculation of *Yersinia pseudotuberculosis* in its right
thigh the lesions are rather similar to those of tuberculosis (**383**). There
are lesions in the swollen local and draining lymph nodes. Many small
grey-white nodules are distributed evenly over the surface of the spleen
and liver, and there are a few in the lungs. (*3 weeks after infection;
reflected light from below, x1.*)

386

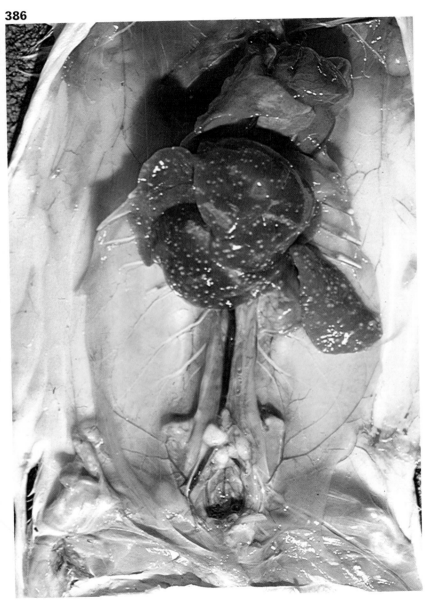

387 *Organs of a guinea-pig naturally infected with Yersinia pseudotuberculosis* The natural portal of entry of this organism is the gut. The most conspicuous lesion is the grossly enlarged mesenteric lymph node, shown surrounded by loops of intestine in the right of the picture. The lesions in the liver and spleen are similar to those shown in **386**, except for a few larger and probably older ones. A comparison of these two figures illustrates how the route of infection determines the distribution and nature of the lesions. (*Museum specimen; reflected light, x¾.*)

388 *Salmonella lesions in the spleen of a guinea-pig* The animal was one of a group of laboratory guinea-pigs in which an outbreak of salmonellosis occurred. Lesions like these might be confused with those of experimentally induced diseases such as tuberculosis. The causative organism should always be identified in the lesions when laboratory animal inoculation is used for the identification of an organism. (*Reflected light, x¾.*)

387

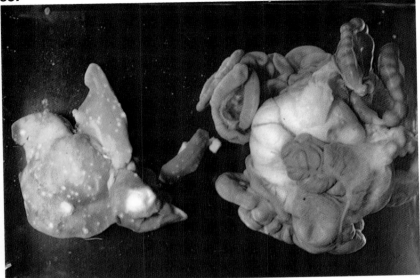

388

389 *Experimental melioidosis in a guinea-pig* The animal was
inoculated intraperitoneally with 0.5ml of 18 hour broth culture of
Pseudomonas pseudomallei. There were a number of abscesses in the
spleen, which is obscured in the photograph by adherent omentum,
which itself contains abscesses. There are other abscesses in the abdominal wall, middle left, and in a lymph node, bottom left. The testes
are anchored to the scrotum by a fibrio-purulent exudate ; this, the
Straus reaction, occurs also after infection with *P. mallei, Actinobacillus
lignieresi, Corynebacterium ovis*, brucellas and a few rarer organisms.
(*5 days after infection; reflected light, x1.*)

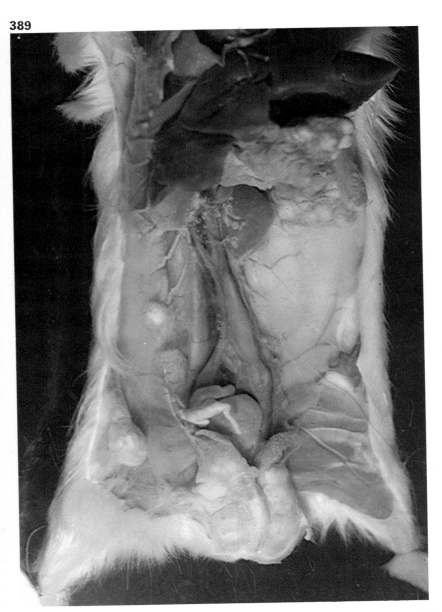

390 *The local lesion resulting from subcutaneous inoculation of Bacillus anthracis into the guinea-pig* There is extensive hae-morrhagic oedema of the subcutaneous tissues. Other characteristic features of this experimental infection which are not shown here are septicaemia and an enlarged, soft spleen. (*The guinea-pig died 2 days after infection; reflected light, x½.*)

391 *Lesion resulting from injection of Clostridium welchii into a guinea-pig* The animal was injected intramuscularly in its right thigh. Within twelve hours it had a crepitant swelling spreading towards the axilla ; the overlying skin was reddened. The incised swelling contained blood-stained oedema ; there were gas pockets in the muscles which were pale and friable. In gas gangrene the nature of the lesion is deter-mined by the infecting clostridia. A mixture of *C. welchii* and *C. sporo-genes* produces a very severe disease with blackening and putrid lique-faction of the tissues, and abundant gas formation (cf. effects of these species on cooked meat, **60**). (*Euthanasia at 18 hours; reflected light, x½.*)

392 *Intradermal reaction to Type A Clostridium welchii toxin*
An injection of 0.2ml of cell-free filtrate was made into the shaved skin of a guinea-pig. After 24 hours the typical lesion of *C. welchii* Type A, toxin is shown – a yellow necrotic area demarcated by a red edge. The lesion has spread vertically downward from the injection site. About midway between this lesion and the right hand in the photograph, an injection was given of the same toxin which had been neutralised by Type A antitoxin. *C. welchii* strains are typed by their intradermal toxin reactions in the guinea-pig, but because of shared toxins a series of neutralisation tests are needed. (*C. welchii grown in Robertson's cooked meat medium for 5 hours at 37°C; lesion photographed 24 hours after injection; reflected light, x¾.*)

390

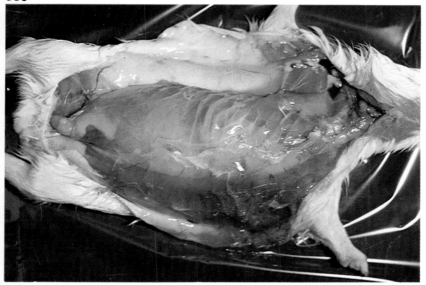

391

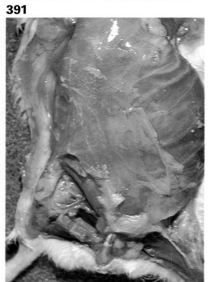

392

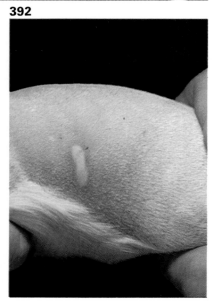

393 *Tetanus in a mouse* An injection of 0.2ml of seven-day old culture of *Clostridium tetani* was made into the subcutaneous tissues at the right of the base of the tail of this mouse. Twenty-four hours later the right hind leg was rigidly extended, as were the individual toes of that leg. The tail was strongly curved to the right. That this resulted from tetanus toxin was confirmed by protection of other mice with specific antitoxin. (*Reflected light, x1.*)

394 *Botulism in a guinea-pig* This guinea-pig was fed eight drops (0.25ml) of a ten day old culture of *Clostridium botulinum*. After eighteen hours it showed flaccid paralysis of the abdomen and neck. It had saliva around the mouth and passed urine and faeces a few minutes before being photographed. At this stage it had shallow breathing accompanied by rales. The only voluntary movements of which it seemed capable were weak twitches of the hind limbs and ears. (*The dose was grown in Robertson's cooked meat medium, 10 days at 37°C; reflected light, x$\frac{1}{3}$.*)

393

394

395 *Experimental listeriosis in a mouse* The animal was inoculated intraperitoneally with 0.005ml of a 24 hour broth culture of *Listeria monocytogenes*. A number of tiny necrotic foci can be seen on the surface of the liver. The lesions in the spleen are larger and more numerous; virtually the whole organ is involved. (*4 days after infection; reflected light, x¾.*)

396 *Experimental infection of a mouse with Streptobacillus moniliformis* An injection of 0.02ml of 18 hour serum broth culture was made into the animal's right hind footpad. A local lesion developed which spread up the limb and eventually involved the tibio-tarsal (hock) joint. Intraperitoneal infection of the mouse may also produce arthritis, provided the animal survives the initial phase of septicaemia. Most commonly the joints of the limbs or tail are involved. (*2 weeks after infection; reflected light, x¾.*)

395

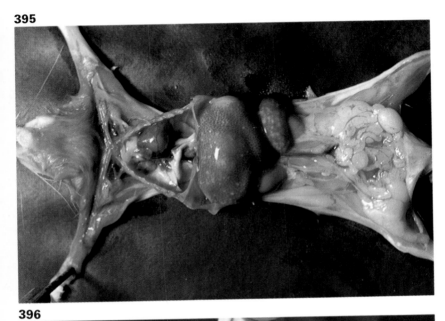

396

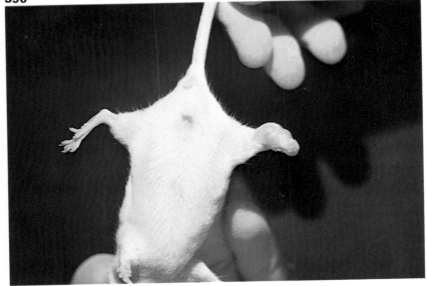

397 Kidney lesions in a mouse following intravenous infection with Candida albicans The animal died 3 days after receiving 0.2ml of 1% suspension of yeast cells. Many abscesses are distributed through-out the kidney cortex. *C. tropicalis* is the only other candida which may produce lesions like these in mice. Of the two only *C. albicans* is patho-genic for the rabbit, which develops similar lesions. (*Reflected light, x1 ½.*)

398 Orchitis in a guinea-pig caused by Paracoccidioides brasiliensis A suspension of a yeast phase culture was injected into the animal's left testicle. The guinea-pig was killed three weeks later. The photograph shows a large volume of pus within the testis. This pus is especially useful for determining the characteristic cellular morphology of the organism (**245**). Mycelial phase cultures produce similar lesions. (*Reflected light, x¾.*)

397

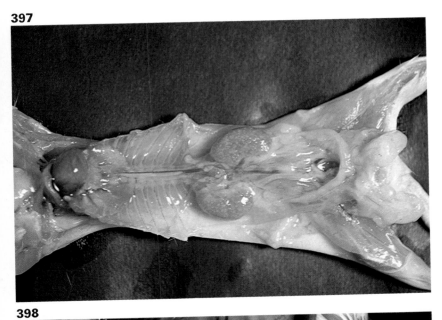

398

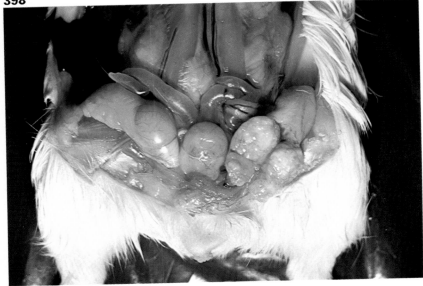

279

References

Anderson, J. S., Happold, F. C., McLeod, J. W., and Thomson, J. G. (1931). On the existence of two forms of diphtheria bacillus — *B. diphtheriae gravis* and *B. diphtheriae mitis* — and a new medium for their differentiation and for the bacteriological diagnosis of diphtheria. Journal of Pathology and Bacteriology, 34, 667.

Bühlmann, X., Vischer, W. A. and Bruhin, H. (1961). Identification of apyocyanogenic strains of *Pseudomonas aeruginosa*. Journal of Bacteriology, 82, 787.

Collins, C. H. and Lyne, P. M. (1970). Microbiological methods, 3rd Edn. London. Butterworths.

Cowan, S. T. and Steel, K. J. (1965). Manual for the identification of medical bacteria. Cambridge. University Press.

Crowle, A. J. (1961). Immunodiffusion. New York & London. Academic Press.

Cruickshank, R. (1969). Medical Microbiology. 11th ed. Edinburgh & London. Livingstone.

Gershman, M. (1963). Modified motility-sulphide medium. Journal of Bacteriology, 86, 1122.

Gray, M. L. and Killinger, A. H. (1966). *Listeria monocytogenes* and listeric infections. Bacteriological Reviews, 30, 309.

Gridley, M. F. (1953). A stain for fungi in tissue sections. American Journal of Clinical Pathology, 23, 303.

Lacey, B. W. (1954). A new selective medium for *Haemophilus pertussis,* containing a diamidine, sodium fluoride and penicillin. Journal of Hygiene, 52, 273.

Lautrop, H. (1960). Laboratory diagnosis of whooping-cough or *Bordetella* infections. Bulletin of the World Health Organisation, 23, 15.

Mackie, T. J. and McCartney, J. E. (1953). Handbook of practical bacteriology, 9th ed. Edinburgh and London. Livingstone.

Marks, J. (1972). Classification of mycobacteria in relation to clinical significance. Tubercle, 53, 259.

Preston, N. W. and Maitland, H. B. (1952). The influence of temperature on the motility of *Pasteurella pseudotuberculosis*. Journal of General Microbiology, 7, 117.

Preston, N. W. and Morrell, A. (1962). Reproducible results with the Gram stain. Journal of Pathology and Bacteriology, 84, 241.

Walker, P. D., Battey, I. and Thomson, R. O. (1971). The localisation of bacterial antigens by the use of the fluorescent and ferritin labelled antibody techniques. In J. R. Norris and D. W. Ribbons, *Methods in Microbiology,* vol 5A. London & New York. Academic Press, p. 219.

Wheeler, E. A., Hamilton, E. G. and Harman, D. J. (1965). An improved technique for the histopathological diagnosis and classification of leprosy. Leprosy Review, 36, 37.

Whittlestone, P. (1969). Isolation of *Mycoplasma suipneumoniae.* In D. A. Shapton and G. W. Gould, *Isolation methods for microbiologists.* London. Academic Press, p. 51.

Index

(The references printed in **bold** *type are to picture and caption numbers, those in light type are to page numbers. An asterisk indicates that the entry is one of the synonyms listed on page 7, and that an alternative name is used in the caption.)*